AF880736

World Atlas
of
Central Solar Eclipses

Color Edition

Fred Espenak

Central Solar Eclipses from 1501 to 2400

Edition 1.0
January 2025

World Atlas of Central Solar Eclipses

Copyright © 2025 by Fred Espenak

All rights reserved. No part of this publication may be reproduced, distributed, or transmitted in any form or by any means, including photocopying, recording, or other electronic or mechanical methods, without the prior written permission of the publisher, except in the case of brief quotations embodied in critical reviews and certain other noncommercial uses permitted by copyright law. For permission requests, write to the publisher at the address below.

Astropixels Publishing
P.O. Box 16197
Portal, AZ 85632

Astropixels Publishing Website: *astropixels.com/pubs*

For more information about this book, see: *astropixels.com/pubs/WACSE.html*

Printed in the United States of America

ISBN 978-1-941983-51-5

Astropixels Publication: AP050 (Version 1.0a)

First Edition

Front Cover: Three maps illustrate the three geographic regions covered by this work:
A) North & South America, B) Europe & Africa, and C) Asia & Australia

Back Cover Photo of Fred Espenak: Copyright © 2020 by Patricia Espenak.

Table of Contents

Photo 1 - The Diamond Ring Effect: In the final seconds before the total eclipse begins, the shrinking crescent Sun produces a short, brilliant segment along one edge of the Moon (the diamond) as the solar corona (the ring) emerges into view in the rapidly darkening sky. 2019 July 2, Vicuna, Chile, ©2019 F. Espenak

Section 1: Solar Eclipse Fundamentals

1.1 Introduction

A total eclipse of the Sun is arguably the most spectacular astronomical event visible to the naked eye. For a few brief minutes, the Sun's brilliant disk is hidden from view and the faint solar corona surrounding the Sun is revealed. The startling event is accompanied by an eerie twilight in which brighter stars and planets are visible in the daytime sky, and the colors of sunset surround the entire horizon.

Although a total solar eclipse occurs once every 12 to 18 months, it is only visible from within the narrow track of the Moon's umbral shadow as it races across Earth's surface. On average, this track or path of totality is only 225 kilometers (140 miles) wide encompassing but a tiny fraction of Earth's surface. It is for this reason that total eclipses are so rare from any one location.

1.2 Classification of Solar Eclipses

There are four basic types of solar eclipses:

1. **Partial Solar Eclipse** – The Moon's penumbral shadow traverses Earth. The Moon's umbral and antumbral shadows completely miss Earth. A portion of the Sun's disk is obscured from within the penumbra.
2. **Annular Solar Eclipse** – The Moon's penumbral and antumbral shadows traverse Earth. The Moon's umbral shadow completely misses Earth. The Moon's disk appears smaller than the Sun so a bright ring surrounds the Moon when viewed from within the antumbral shadow. A partial eclipse is seen within the penumbral shadow.
3. **Total Solar Eclipse** – The Moon's penumbral and umbral shadows traverse Earth. The Moon's antumbral shadow extends beyond Earth's surface. The Moon's disk appears larger than the Sun and completely covers the solar disk when viewed from within the umbral shadow. A partial eclipse is seen within the penumbral shadow.
4. **Hybrid Solar Eclipse** – The Moon's penumbral, umbral and antumbral shadows all traverse different parts of Earth. The curvature of Earth's surface brings some regions into the umbra and others into the antumbra. The eclipse appears total within the umbra and annular within the antumbra. A partial eclipse is seen within the penumbral shadow. Hybrid eclipses are also known as annular-total eclipses.

Figure 1–1: Geometry of an Annular Solar Eclipse

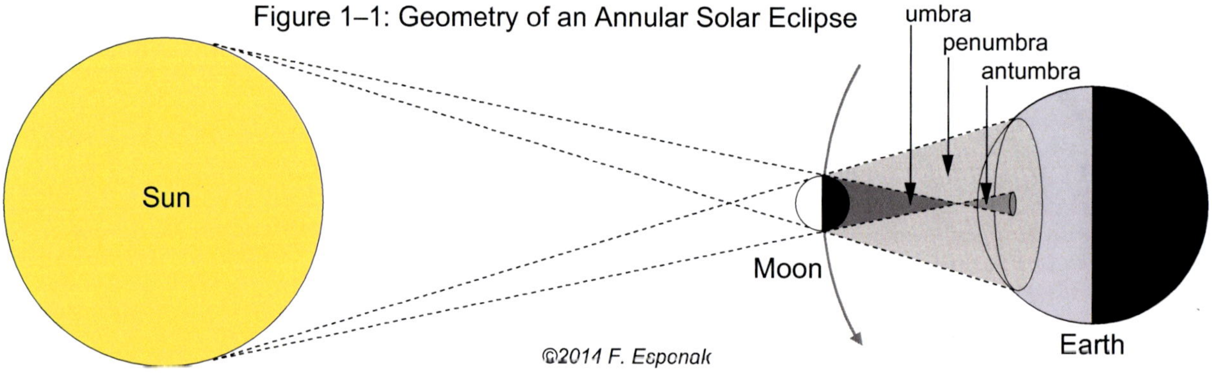

Figure 1–1 illustrates the geometry of an annular solar eclipse. A partial eclipse is visible from within the large penumbral shadow, while the annular eclipse is confined to the much smaller antumbral shadow.

Figure 1–2: Geometry of a Total Solar Eclipse

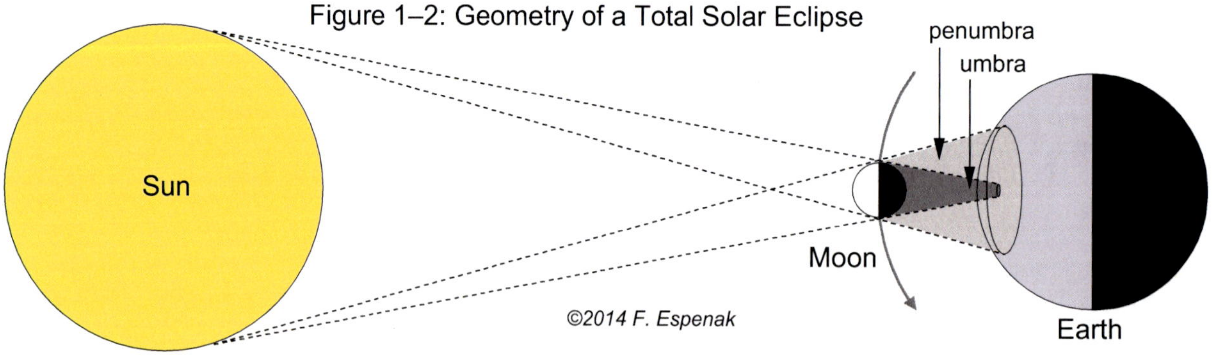

Figure 1–2 illustrates the geometry of a total solar eclipse. A partial eclipse is visible from within the large penumbral shadow, while the total eclipse is only seen from the much smaller umbral shadow.

1.3 Central Solar Eclipses

When the central axis of the Moon's umbral or antumbral shadow strikes Earth, the resulting eclipse is either a total, annular or hybrid. Thus, the term central[1] eclipse is a convenient way of grouping these three eclipses into a single category. In contrast, the central shadow axis always misses Earth during a partial eclipse.

The Moon casts a second shadow called the penumbra that is approximately 6700 to 7300 kilometers (4200 to 4500 miles) in diameter and can cover a significant fraction of the daytime hemisphere of Earth. Anyone inside the penumbra will observe a partial solar eclipse. Consequently, all central eclipses begin and end with a partial eclipse visible from large geographic areas as the penumbra sweeps across Earth's surface.

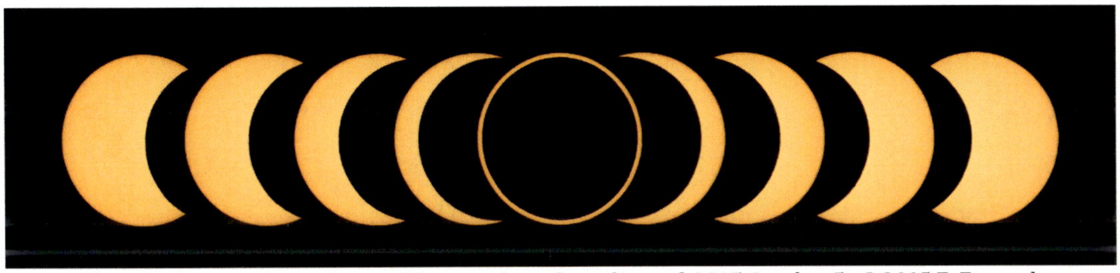

Photo 2 shows various phases of the Annular solar eclipse of 2005 October 5. ©2005 F. Espenak

1.4 Visual Appearance of Annular Solar Eclipses

During an annular eclipse, the Moon's penumbral and antumbral shadows sweep across Earth. While the penumbra is quite large, the antumbra is much smaller and has a maximum diameter of 374 kilometers (232 miles). Because of this, the antumbra covers a much smaller fraction of Earth's surface.

The Moon's orbital motion carries the penumbral and antumbral shadows in a west to east direction. A partial eclipse is visible within the penumbra, but only observers located in the much narrower path of the antumbra will see an annular eclipse. The antumbra's track across Earth is called the path of annularity.

All annular eclipses begin with a series of partial phases lasting up to an hour or more. At the peak of the eclipse, the Moon's disk can be seen in complete silhouette against the Sun. The remaining solar photosphere appears as an intensely bright ring of light surrounding the Moon. The annular phase lasts a maximum of 12 ½ minutes, but it is more typically 3 to 6 minutes in length. After annularity, another series of partial phases occur as the Moon gradually uncovers the Sun.

Even during the annular phase, the Sun is dangerously bright and cannot be viewed without a solar filter. In this regard, annular eclipses are similar to partial eclipses. The landscape and sky remain bright throughout the eclipse, giving little indication of the celestial event in progress.

1.5 Visual Appearance of Total Solar Eclipses

During a total eclipse, the Moon's penumbral and umbral shadows fall upon Earth. The umbra is much smaller than the penumbra and has a maximum diameter of 273 kilometers (170 miles). The narrow track traced out by the umbra as it sweeps west to east across Earth's surface is called the path of totality. Anyone standing within this zone will see the Sun completely obscured by the Moon. The total phase can last up to 7 ½ minutes, but is more typically 2 to 3 minutes in length.

---

[1] Under rare circumstances, it is possible to have a total or annular eclipse that is not central. Such events occur at high latitudes when the umbral or antumbral shadow just grazes Earth while the central axis misses. Such events are classified as non-central total (0.9%) or non-central annular eclipses (1.7%).

Total eclipses all begin and end with a series of partial phases lasting up to an hour or more. But this is where the resemblance between partial and annular eclipses ends. The total phase is the most spectacular astronomical event visible to the naked eye. At this time, the Sun's outer atmosphere — the solar corona — appears as a gossamer halo surrounding the Moon, and bright stars and planets are visible.

Photo 3 captures details in the solar corona during the total eclipse of 2006 March 29. ©2006 F. Espenak

The eclipse starts to take on a unique character about five minutes before the total phase begins. Sunlight has a foreboding quality and casts abnormally sharp shadows. The approaching lunar umbra darkens the western sky and the air temperature is noticeably cooler. A minute before totality, ghostly shadow bands[2] ripple across the ground. The ambient light grows feeble even though the crescent Sun is still too bright to look at. In the final seconds, the Sun's corona emerges from the glare as the solar crescent shrinks to a single brilliant jewel. This celestial diamond ring lingers for a moment before the last bead of sunlight is extinguished.

The Sun's glorious corona is now displayed to full advantage in the darkened sky of totality. The simple act of standing within the shadow of the Moon affords the rare opportunity to gaze directly at the glowing million-degree plasma surrounding our star. Twisted and constrained by the Sun's enormous magnetic fields, the solar corona is revealed to the naked eye only during the brief seconds when the Moon completely blocks the Sun's brilliant disk. An eerie twilight bathes the landscape and the colors of dusk surround the horizon.

The minutes race by like seconds. Suddenly, a sparkling bead of sunlight reappears along one edge of the Moon and quickly grows to blindingly bright proportions. Daylight returns as the corona fades and the total phase ends. Although another hour of partial phases remains before the eclipse ends, these are anticlimactic after the spectacle of totality.

While filters are required for viewing the partial phases, they must be removed for totality. The total phase is the only time it is completely safe to look directly at the Sun without protection. In fact, the total phase is not even visible through solar filters because the Sun's corona is a million times fainter than the photosphere.

---

[2] Shadow bands are often seen immediately before and after a total solar eclipse. They appear as thin parallel wavy lines of alternating light and dark that race across the ground.

Section 2: Solar Eclipse Statistics

2.1 Introduction

During the 900-year period from 1501 to 2400, Earth experiences 2,152 eclipses of the Sun. The statistical distribution of the four eclipse types over this interval is shown in Table 2-1.

Table 2-1: Distribution of Basic Eclipse Types

Eclipse Type	Abbreviation	Number	Percent
All Eclipses	—	2152	100.0%
Partial	P	757	35.2%
Annular	A	705	33.2%
Total	T	585	27.2%
Hybrid	H	105	4.5%

All partial eclipses are events in which some portion of the Moon's penumbral shadow passes across Earth's surface. In comparison all annular, total and hybrid eclipses can be characterized as events in which some portion of the Moon's umbral and/or antumbral shadow crosses Earth.

In the case of umbral or antumbral eclipses (annular, total or hybrid), they can be further categorized as:

1. **Central (two limits)** — The central axis of the Moon's umbral or antumbral shadow traverses Earth thereby producing a central line in the eclipse track. The umbra or antumbra falls entirely upon Earth producing a ground track with both a northern and southern limit.
2. **Central (one limit)** — The central axis of the Moon's umbral or antumbral shadow traverses Earth. However, a portion of the umbra or antumbra misses Earth throughout the eclipse thereby producing a ground track with just one limit.
3. **Non-Central** — The central axis of the Moon's umbral or antumbral shadow misses Earth. However, one edge of the umbra or antumbra grazes Earth thereby producing a ground track with one limit and no central line.

Using these categories, the distribution of the 705 annular eclipses appears in Table 2-2.

Table 2-2: Statistics of Annular Eclipses

Annular Eclipses	Number	Percent
All Annular Eclipses	705	100.0%
Central (two limits)	682	96.7%
Central (one limit)	11	1.6%
Non-Central (one limit)	12	1.7%

Examples of central annular eclipses with one limit include: 1874 Oct 10, 2003 May 31, 2044 Feb 28, and 2101 Feb 28. Some examples of non-central annular eclipses are: 1950 Mar 18, 1957 Apr 30, 2014 Apr 29, and 2043 Oct 03.

Similarly, the distribution of the 585 total eclipses is shown in Table 2-3.

Table 2-3: Statistics of Total Eclipses

Total Eclipses	Number	Percent
All Total Eclipses	585	100.0%
Central (two limits)	576	98.5%
Central (one limit)	4	0.7%
Non-Central (one limit)	5	0.9%

Examples of central total eclipses with one limit include: 1523 Aug 11, 2185 Jul 26, and 2195 Aug 05. Several examples of non-central total eclipses are: 1957 Oct 23, 1967 Nov 02, and 2043 Apr 09.

All 105 hybrid eclipses are central with two limits. Hybrid eclipses with a single limit (both central and non-central) are exceedingly rare. An estimate of the mean frequency of non-central hybrid eclipses is one out of every 600 million eclipses or once every 250 million years (Meeus, 2002).

The central path of most hybrid eclipses begins annular, changes to total and reverts back to annular. This combination (ATA) occurs in 99 out of the 105 hybrid eclipses in the *World Atlas*. However, there are two other possibilities. If the vertex of the Moon's umbral shadow passes through Earth's fundamental plane during the eclipse, then the hybrid eclipse can begin as total and end as annular (TA) or it can begin as annular and end as total (AT). Table 2–4 shows the distribution of the three classes of hybrid eclipses.

Table 2–4: Statistics of Hybrid Eclipses

Hybrid Eclipses	Number	Percent
All Hybrid Eclipses	105	100.0%
Hybrid (ATA)	98	93.3%
Hybrid (TA)	4	3.8%
Hybrid (AT)	3	2.8%

Examples of ATA hybrid eclipses include: 1986 Oct 03, 1987 Mar 29, 2005 Apr 08, and 2023 Apr 20. Examples of the relatively rare TA hybrid eclipse are: 1564 Jun 08, 1703 Jan 17, 1825 Dec 09, and 2386 Apr 29. Finally, examples of the rare AT hybrid eclipse include: 1854 Nov 20, 2013 Nov 03, and 2172 Oct 17.

2.2 Distribution of Eclipse Types by Century

Table 2-5 summarizes 900 years of eclipses by eclipse type in 100-year intervals. The number of central and non-central (in square brackets) events is given for annular and total eclipses. The number of eclipses in any one century ranges from 224 to 251 with an average of 239.1.

Table 2-5: Eclipse Types by Century: 1501 to 2400

Century Interval	Number of Eclipses	Number of Partial Eclipses	Number of Annular Eclipses*	Number of Total Eclipses*	Number of Hybrid Eclipses
1501 to 1600	228	75	69 [3]	62 [0]	19
1601 to 1700	248	89	74 [0]	60 [1]	24
1701 to 1800	251	92	78 [0]	62 [0]	19
1801 to 1900	242	87	77 [0]	63 [0]	15
1901 to 2000	228	78	71 [2]	68 [3]	6
2001 to 2100	224	77	70 [2]	67 [1]	7
2101 to 2200	235	79	82 [5]	65 [0]	4
2201 to 2300	248	92	86 [0]	67 [0]	3
2301 to 2400	248	88	86 [0]	66 [0]	8

* The first quantity is the number of central eclipses, while the second quantity in square brackets is the number of non-central eclipses.

Some remarkable patterns are present in this table. There is a cyclical variation in the number of eclipses per century with a period just under six centuries, giving alternating "rich" and "poor" periods (Meeus, 1997). The 20th and 21st centuries (1901 to 2100) are poor periods, with only 228 and 224 eclipses. This cycle is also present when only central eclipses are considered.

The cycle appears to have a period of approximately 600 years with an amplitude of ~30 eclipses. This is close to a known lunar eclipse period called the tetradia, which has a period of 586.02 years. The tetradia governs the recurrence of tetrads or groups of four successive total lunar eclipses each separated by six lunations. The tetradia cycle for lunar eclipse tetrads appears to be 180 degrees out of phase with the cycle for solar eclipses. When there are many tetrads, there are fewer solar eclipses. We are currently in a tetrad rich period with contemporary tetrads in 2003–2004, 2014–2015 and 2032–2033.

The number of hybrid solar eclipses per century also varies cyclically with a period of ~17 centuries.

2.3 Extremes in Eclipse Magnitude: Annular Eclipses

Six annular eclipses have a maximum magnitude at greatest eclipse less than or equal to 0.910 (Table 2-6). Most of these events are central, but the last one is non-central with one limit.

Table 2-6: Annular Eclipses with Magnitude 0.910 or Less

Date (Dynamical Time)	Saros	Gamma	Eclipse Magnitude	Central Duration
1565 Nov 22	135	0.9564	0.9092	09m 37s
1583 Dec 14	135	0.9471	0.9083	10m 03s
1601 Dec 24	135	0.9402	0.9078	10m 14s
1620 Jan 04	135	0.9322	0.9081	10m 13s
1638 Jan 15*	135	0.9242	0.9090	10m 00s

* Non-central annular eclipse with one limit.

Five annular eclipses have a maximum magnitude at greatest eclipse greater than or equal to 0.9995 (Table 2-7). All of these events have central durations lasting 2 seconds or less.

Table 2-7: Annular Eclipses with Magnitude 0.9995 or More

Date (Dynamical Time)	Saros	Gamma	Eclipse Magnitude	Central Duration
1704 Nov 27	118	0.6716	0.9999	00m 01s
1822 Feb 21	137	0.6914	0.9996	00m 02s
1858 Mar 15	137	0.6461	0.9996	00m 02s
1876 Mar 25	137	0.6142	0.9999	00m 01s
1948 May 09	137	0.4133	0.9999	00m 00s

2.4 Extremes in Eclipse Magnitude: Total Eclipses

Seven total eclipses have a magnitude less than 1.01 (Table 2-8). Three of them are non-central eclipses.

Table 2-8: Total Eclipses with Magnitude 1.01 or Less

Date (Dynamical Time)	Saros	Gamma	Eclipse Magnitude	Central Duration
1522 Mar 27	131	0.9947	1.0076	00m26s
1536 Dec 13	134	-0.9343	1.0098	00m33s
1554 Dec 24	134	-0.9341	1.0075	00m25s
1562 Feb 03	133	0.9373	1.0091	00m41s
1957 Oct 23	123	-1.0022	1.0013	*
1968 Sep 22	124	0.9451	1.0099	00m40s
2043 Apr 09	149	1.0031	1.0096	*

* These eclipses are non-central, so there are no central lines nor central durations.

Eleven total eclipses have a maximum magnitude of 1.079 or more. The central durations of nearly all exceed six minutes. They take place when Earth is near aphelion (June–August), resulting in a small solar disk.

Table 2-9: Total Eclipses with Magnitude 1.079 or More

Date (Dynamical Time)	Saros	Gamma	Eclipse Magnitude	Central Duration
1973 Jun 30	136	-0.0785	1.0792	07m04s
1991 Jul 11	136	-0.0041	1.0800	06m53s
2009 Jul 22	136	0.0698	1.0799	06m39s
2027 Aug 02	136	0.1421	1.0790	06m23s
2150 Jun 25	139	-0.0911	1.0802	07m14s
2168 Jul 05	139	-0.1660	1.0807	07m26s
2186 Jul 16	139	-0.2396	1.0805	07m29s
2204 Jul 27	139	-0.3129	1.0793	07m22s
2327 Jun 20	142	0.2542	1.0795	06m21s
2345 Jun 30	142	0.3267	1.0797	06m07s
2363 Jul 12	142	0.4012	1.0792	05m51s

2.5 Extremes in Eclipse Magnitude: Hybrid Eclipses

Six hybrid eclipses have a maximum magnitude at greatest eclipse less than or equal to 1.0002. All of these events are central with a central duration of totality of 1 second or less.

Table 2-10: Hybrid Eclipses with Magnitude 1.0002 or Less

Date (Dynamical Time)	Saros	Gamma	Eclipse Magnitude	Central Duration
1612 Nov 22	136	-0.7692	1.0002	00m 01s
1627 Aug 11	139	0.9401	1.0001	00m 00s
1702 Jul 24	131	0.3160	1.0002	00m 01s
1804 Feb 11	137	0.7053	1.0001	00m 00s
1894 Apr 06	137	0.5740	1.0001	00m 01s
1986 Oct 03	124	0.9931	1.0000	00m 00s

Eight hybrid eclipses have a maximum magnitude at greatest eclipse greater than or equal to 1.014. Six events have a duration of totality of 1 minute 30 seconds or more.

Table 2-11: Hybrid Eclipses with Magnitude 1.014 or More

Date (Dynamical Time)	Saros	Gamma	Eclipse Magnitude	Central Duration
1564 Jun 08	120	0.1252	1.0174	01m44s
1705 May 22	123	-0.1525	1.0147	01m32s
1825 Dec 09	139	0.5296	1.0148	01m34s
1854 Nov 20	140	-0.5179	1.0144	01m07s
1864 May 06	126	0.2622	1.0146	01m25s
2013 Nov 03	143	0.3272	1.0159	01m40s
2172 Oct 17	146	-0.1484	1.0174	01m34s
2386 Apr 29	154	-0.5483	1.0147	01m30s

2.6 Greatest Central Duration: Annular Eclipses

Nine annular eclipses have a central duration at greatest eclipse of 11 1/2 minutes or more. They all take place near Earth's perihelion (Dec–Jan), resulting in a large solar disk.

Table 2-12: Annular Eclipses with Central Duration of 11½ Minutes or More

Date (Dynamical Time)	Saros	Gamma	Eclipse Magnitude	Central Line Duration
1592 Dec 03	116	0.6102	0.9159	11m36s
1610 Dec 15	116	0.6195	0.9153	11m56s
1628 Dec 25	116	0.6264	0.9153	12m02s
1647 Jan 05	116	0.6336	0.9161	11m50s
1919 Nov 22	141	0.4549	0.9198	11m37s
1937 Dec 02	141	0.4389	0.9184	12m00s
1955 Dec 14	141	0.4266	0.9176	12m09s
1973 Dec 24	141	0.4171	0.9174	12m02s
1992 Jan 04	141	0.4091	0.9179	11m41s

2.7 Greatest Central Duration: Total Eclipses

Eight total eclipses have a central duration at greatest eclipse of 7 minutes or more. They take place when Earth is near aphelion (June–August), resulting in a small solar disk. The longest duration of totality occurs on 2186 Jul 16. Its central duration of 7 minutes 29 seconds is close to the theoretical maximum of 7 minutes 32.1 seconds during that epoch (Meeus, 2004). The eight eclipses are members of just two Saros series. The eclipses of 1937, 1955 and 1973 all belong to Saros 136. This is the same Saros producing the 6+ minute eclipses in 1991, 2009 and 2027. In all cases the eclipse magnitude is greater than 1.07.

Table 2-13: Total Eclipses with Central Duration of 7 Minutes or More

Date (Dynamical Time)	Saros	Gamma	Eclipse Magnitude	Central Duration
1937 Jun 08	136	-0.225	1.075	07m 04s
1955 Jun 20	136	-0.153	1.078	07m 08s
1973 Jun 30	136	-0.079	1.079	07m 04s
2150 Jun 25	139	-0.091	1.080	07m 14s
2168 Jul 05	139	-0.166	1.081	07m 26s
2186 Jul 16	139	-0.240	1.080	07m 29s
2204 Jul 27	139	-0.313	1.079	07m 22s
2222 Aug 08	139	-0.384	1.077	07m 06s

2.8 Greatest Central Duration: Hybrid Eclipses

Six hybrid eclipses have a central duration at greatest eclipse greater than or equal to 1 minute 30 seconds. . In all cases the eclipse magnitude is greater than 1.014

Table 2-14: Hybrid Eclipses with Central Duration of 1m 30s or More

Date (Dynamical Time)	Saros	Gamma	Eclipse Magnitude	Central Duration
1564 Jun 08	120	0.1252	1.0174	01m44s
1705 May 22	123	-0.1525	1.0147	01m32s
1825 Dec 09	139	0.5296	1.0148	01m34s
2013 Nov 03	143	0.3272	1.0159	01m40s
2172 Oct 17	146	-0.1484	1.0174	01m34s
2386 Apr 29	154	-0.5483	1.0147	01m30s

Section 3: Explanation of Atlas Eclipse Maps: Appendices A, B & C

3.1 Introduction

Earth experiences 1395 central eclipses (annular, total and hybrid eclipses)[3] of the Sun during the 900-year period from 1501 to 2400. That averages to 155 central eclipses per century.

The 81 full-page maps in Appendices A, B and C show the path of every central solar eclipse (annular, total and hybrid) with respect to continental coastlines, political boundaries (circa 2018), and major cities. The maps employ a transverse Mercator projection[4] and are centered on three geographic regions:

Appendix A – North and South America
Appendix B – Europe and Africa
Appendix C – Asia and Australia

There are three maps per century for each Appendix The maps cover periods of 33-years, 33-years and 34-years in a given century. For instance, the 16th century is covered by three maps in each Appendix:

1. 1501–1533
2. 1534–1566
3. 1567–1600

On the maps, each eclipse is identified by its date at the instant of greatest eclipse. An asterisk symbol marks the location of greatest eclipse (if that point of the eclipse path is present on a given map).

The paths of total, annular, and hybrid eclipses can be identified as follows.

Total Eclipses –
1. the path is shaded blue
2. the date appears in bold and is preceded by the letter **T**

Annular Eclipses –
1. the path is shaded orange
2. the date appears in italics and is preceded by the letter **A**

Hybrid Eclipses –
1. the path is shaded blue along the total sections and shaded orange along the annular sections
2. the date appears in bold-italics and is preceded by the letter **H**

---

[3] For the purposes of this publication, we include the 12 non-central annular and 5 non-central total eclipses in the overall number of 1395 central eclipse from 1501 to 2400.

[4] The transverse Mercator projection (also known as the Gauss-Krüger projection) is similar to Mercator except that the cylinder touches the sphere or ellipsoid along a meridian instead of the equator. The result is a conformal projection that correctly shows the shape of features but does not maintain true directions.

Section 4: Explanation of Central Solar Eclipse Catalog: Appendix D

4.1 Introduction

Earth experiences 1395 central eclipses (annular, total and hybrid eclipses)[5] of the Sun during the 900 year period from 1501 to 2400. The catalog in Appendix D consists of a series of tables that summarize the principal characteristics of each solar eclipse over this time interval. The tables compliment the eclipse maps in Appendices A, B, and C.

Each line in the catalog corresponds to one eclipse and provides concise parameters to characterize the eclipse. The calendar date and Dynamical Time of the instant of greatest eclipse are given along with the adopted value of delta T [ΔT] (Section 5.4). The lunation number and the Saros series are listed along with the eclipse type. Gamma is the distance of the shadow axis from Earth's center at greatest eclipse, while the eclipse magnitude is defined as the fraction of the Sun's diameter obscured at that instant. The geographic latitude and longitude of the umbra are given for greatest eclipse, along with the Sun's altitude and azimuth, the width of the path, and the central line duration of totality or annularity.

4.2 Calendar Date

The Julian calendar is used prior to 1582 Oct 15. All eclipse dates from 1582 Oct 15 onwards use the modern Gregorian calendar currently found throughout most of the world. Because of the Gregorian Calendar Reform, the day following 1582 Oct 04 [Julian calendar] is 1582 Oct 15 [Gregorian calendar] (Section 5.5).

4.3 TD of Greatest Eclipse (Terrestrial Dynamical Time of Greatest Eclipse)

The instant of greatest eclipse occurs when the distance between the axis of the Moon's shadow cone and the center of Earth reaches a minimum. For partial eclipses, the instant of greatest eclipse differs slightly from the instant of greatest magnitude due to Earth's flattening. For total eclipses, the instant of greatest eclipse differs slightly from the instant of greatest duration, although the differences are quite small.

Greatest eclipse is given in Terrestrial Dynamical Time or TD (Section 5.3), which is a time system based on International Atomic Time. As such, TD is the modern equivalent to its predecessor Ephemeris Time and is used in the theories of planetary motion in the Solar System. To determine the geographic visibility of an eclipse, TD is converted to Universal Time (UT1) using the parameter ΔT (Section 5.4).

4.4 ΔT (Delta T)

Delta T (ΔT) is the arithmetic difference, in seconds, between Terrestrial Dynamical Time (TD) and Universal Time (UT1) (Section 5.4).

4.5 Saros Num (Saros Series Number)

Each eclipse belongs to a Saros series using a numbering system first introduced by van den Bergh (1955). The eclipses with an odd Saros number take place at the ascending node of the Moon's orbit; those with an even Saros number take place at the descending node.

The Saros is a period of 223 synodic months, or approximately 18 years, 11 days, and 8 hours. Eclipses separated by this period belong to the same Saros series and share similar geometry and characteristics.

---

[5] For the purposes of this publication, we include the 12 non-central annular and 5 non-central total eclipses in the overall number of 1395 central eclipse from 1501 to 2400.

4.6 Ecl. Type (Solar Eclipse Type)

The first character in this 2-character string gives the eclipse type. The four basic types of solar eclipses are:

1. A = Annular Solar Eclipse – The Moon's antumbral shadow traverses Earth;
   The Moon is too far from Earth to completely cover the Sun
2. T = Total Solar Eclipse – The Moon's umbral shadow traverses Earth;
   The Moon is close enough to Earth to completely cover the Sun
3. H = Hybrid Solar Eclipse – The Moon's umbra and antumbra traverse different parts of Earth;
   Eclipse appears either total or annular along different sections of its path

The second character of the eclipse type is a qualifier defined as follows.

1. m = Middle eclipse of Saros series
2. n = Central eclipse with no northern limit
2. s = Central eclipse with no southern limit
3. + = Non-central eclipse with no northern limit
4. – = Non-central eclipse with no southern limit
5. 2 = Hybrid eclipse path begins total and ends annular
6. 3 = Hybrid eclipse path begins annular and ends total

4.7 Gamma

Gamma is the minimum distance from the axis of the lunar shadow cone to the center of Earth, in units of Earth's equatorial radius. This distance is positive or negative, depending on whether the axis of the shadow cone passes north or south of Earth's center. If gamma is between +0.997 and –0.997, the eclipse is central (either total, annular, or hybrid). The limiting value 0.997 differs from unity because of the flattening of Earth.

4.8 Ecl. Mag. (Eclipse Magnitude)

The eclipse magnitude is defined as the fraction of the Sun's diameter occulted by the Moon. For central eclipses (total, annular, and hybrid), the eclipse magnitude listed is actually the ratio of the topocentric apparent diameters of the Moon and Sun at greatest eclipse. The eclipse magnitude is always less than 1.0 for partial and annular eclipses, but equal to, or greater than, 1.0 for total and hybrid eclipses.

4.9 Lat. Long. (Latitude and Longitude)

The latitude and longitude corresponds to the geographic position of greatest eclipse.

4.10 Sun Alt and Sun Azm (Altitude and Azimuth of Sun)

The Sun's altitude and azimuth at the geographic position intersected by the axis of the lunar shadow cone is given at the instant of greatest eclipse. The azimuth values 0°, 90°, 180°, and 270° correspond to the cardinal directions north, east, south and west, respectively.

4.11 Path Width

For central eclipses, the width of the path of totality or annularity (kilometers) is given at the geographic position intersected by the axis of the lunar shadow cone at the instant of greatest eclipse.

4.12 Central Line Dur. (Central Line Duration)

The central line duration of the total or annular phase (in minutes and seconds) is given at the geographic position intersected by the axis of the lunar shadow cone at greatest eclipse. For total and hybrid eclipses, this is the maximum duration of the total phase along the entire umbral path. For annular eclipses, greatest eclipse may be near either the minimum or maximum duration of the annular phase along the path. If the annular phase duration exceeds ~ 2.3 min, then it corresponds to the maximum duration along the central line track. If the annular phase duration is less, however, then it corresponds to a minimum and the annular duration increases towards the ends of the central path.

Section 5: Solar Eclipse Predictions

5.1 Mean Lunar Radius

A parameter used in the prediction of solar eclipses is the Moon's mean radius k, expressed in units of Earth's equatorial radius. This work uses a value of k=0.272281 for all central solar eclipses and represents the mean *minimum* radius of the Moon. This is smaller than the IAU's adopted a value of k=0.2725076 for the mean *average* lunar radius. The difference is significant because the use of a *minimum* radius more accurately discriminates between total and annular eclipses. An eclipse is considered total only when the Sun's disk is completely occulted by the Moon. Using a smaller value of k is more representative of the deepest lunar valleys, hence the minimum solid disk diameter and ensures that an eclipse is truly total.

5.2 Solar and Lunar Coordinates

The coordinates of the Sun and Moon used in the eclipse predictions have been calculated with the JPL DE431 (JPL Developmental Ephemeris 431). The DE431 is based upon the International Celestial Reference Frame (ICRF) version 2.0, the adopted reference frame of the International Astronomical Union (IAU). The DE431 is often referred to as the "JPL long ephemeris" because it covers a 30,391-year period from -13,200 Aug 15 (JED -0.3100015.5) to 17,191 Mar 15 (JED 8000015.5). The DE431 does not include nutation or libration. These small effects are not visible on the scale of the maps in the **World Atlas**.

5.3 Measurement of Time

In 1884, Greenwich Mean Time (GMT) — the mean solar time on the Greenwich Meridian (0° longitude) — was adopted as the standard reference time for clocks around the world. A fundamental basis of GMT is the assumption that Earth's rotation on its axis is constant. By the mid-twentieth century astronomers realized that Earth's rotation is slowing down because of tidal friction with the Moon.

For purposes of orbital calculations, time using Earth's rotation was abandoned for a more uniform time scale based on Earth's orbit about the Sun. In 1952, Ephemeris Time was introduced to address the problem and was used for Solar System ephemeris calculations until 1979.

Terrestrial Dynamical Time (TD) is the modern replacement for Ephemeris Time and is used in the theories of planetary motion in the Solar System. TD is based on International Atomic Time (TAI), which is a high-precision standard using several hundred atomic clocks worldwide. To ensure continuity with Ephemeris Time, TD was defined to match ET for the date 1977 Jan 01.

Civilian time used throughout the world is still based on mean solar time, although indirectly. While Greenwich Mean Time was determined though observations of the Sun, its modern day replacement, Universal Time (actually UT1) is based on Earth's rotation using observations of distant quasars. UT1 is a nonuniform time because Earth is gradually slowing down at an irregular rate.

Coordinated Universal Time (UTC) is derived from International Atomic Time (TAI). UTC was defined to closely parallel UT1. However, the two time systems are intrinsically incompatible since UTC is uniform while UT1 is based on Earth's rotation, which is gradually slowing. In order to keep the two times within 0.9 seconds of each other, a leap second is added to UTC as needed (currently once every few years).

Today, UTC is the time standard used to define time zones around the world. It is the time reference for GPS satellites and aviation, and is used to synchronize the clocks of computers across the Internet.

5.4 ΔT (Delta T)

The orbital positions of the Sun and the Moon, required by eclipse predictions, are calculated using Terrestrial Dynamical Time (TD) because it is a uniform time scale. However, world time zones are based on Universal Time[6] (UT1). In order to convert eclipse predictions from TD to UT1, the difference between these two time scales must be known. The parameter ΔT (delta-T) is the arithmetic difference, in seconds, between the two as:

$$\Delta T = TD - UT1 \qquad (5\text{–}1)$$

Past values of ΔT can be deduced from historical records. In spite of their relatively low precision, these data represent the only evidence for the value of ΔT prior to 1600. In the centuries following the introduction of the telescope (circa 1609), thousands of high quality observations have been made of lunar occultations of stars, affording valuable data with increased accuracy in the determination of ΔT.

The estimated uncertainty in the value of ΔT is 55 seconds in the year 1000, but it drops to 1 second by 1800. A detailed analysis of historical measurements fitted with cubic splines for ΔT from 1000 to +1950 is presented in Morrison and Stephenson, 2004.

In modern times, the determination of ΔT is made using atomic clocks and radio observations of quasars. From 1955 to 2010, the average 1-year change in ΔT ranges from 0.18 seconds to 1.06 seconds. Future changes in ΔT are unknown since theoretical models of the physical causes are imprecise. Extrapolations from the table weighted by the long period trend from tidal braking of the Moon offer estimates of +76 seconds in 2050, and +94 seconds in 2100.

5.5 Calendar Date

The Gregorian calendar is the civil calendar currently used throughout most of the world. The older Julian calendar was used until 1582 Oct 04. As a consequence of the Gregorian Calendar Reform, the day following 1582 Oct 04 (Julian calendar) is 1582 Oct 15 (Gregorian calendar).

Pope Gregory XIII decreed the use of the Gregorian calendar in 1582 in order to correct a problem in a drift of the seasons. It adopts the convention of a year containing 365 days. Every fourth year is a leap year of 366 days if it is divisible by 4 (e.g., 2004, 2008, etc.). However, whole century years (e.g., 1700, 1800, 1900) are excluded from the leap year rule unless they are also divisible by 400 (e.g., 2000). This dating scheme was designed to keep the vernal equinox on or within a day of March 21. Precession of the equinoxes will eventually produce an error of one day in the Gregorian calendar in about 7700 years.

Prior to the Gregorian Calendar Reform of 1582, the Julian calendar was in wide use. It was less complicated, in that all years divisible by 4 were counted as 366-day leap years, but this simplicity came at a cost. After more than 16 centuries of use, the Julian calendar date of the vernal equinox had drifted 11 days from March 21. It was this failure in the Julian calendar that prompted the Gregorian Calendar Reform.

---

[6] World time zones are actually based on Coordinated Universal Time (UTC). It is an atomic time synchronized and adjusted to stay within a second of astronomically determined Universal Time (UT1) through the addition of an occasional "leap second" to compensate for the gradual slowing of Earth's rotation.

Appendix A

World Atlas of Central Solar Eclipses

North & South America: 1501 to 2400

Index for Appendix A

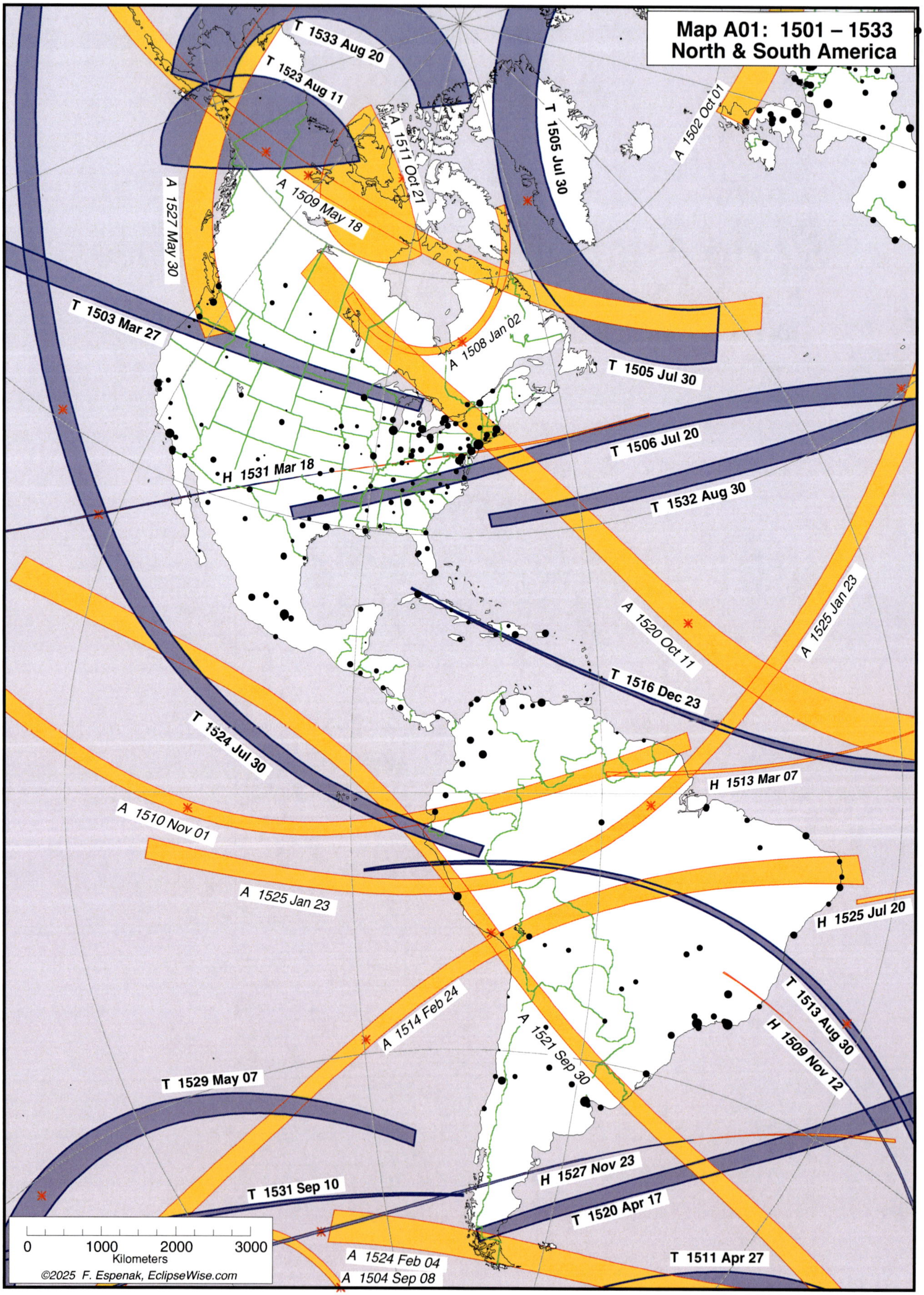
Map A01: 1501 – 1533
North & South America
T 1533 Aug 20
T 1523 Aug 11
A 1511 Oct 21
T 1505 Jul 30
A 1502 Oct 01
A 1527 May 30
A 1509 May 18
T 1503 Mar 27
A 1508 Jan 02
T 1505 Jul 30
T 1506 Jul 20
H 1531 Mar 18
T 1532 Aug 30
A 1520 Oct 11
A 1525 Jan 23
T 1516 Dec 23
T 1524 Jul 30
H 1513 Mar 07
A 1510 Nov 01
A 1525 Jan 23
H 1525 Jul 20
T 1513 Aug 30
H 1509 Nov 12
A 1514 Feb 24
A 1521 Sep 30
T 1529 May 07
H 1527 Nov 23
T 1531 Sep 10
T 1520 Apr 17
0 1000 2000 3000
Kilometers
©2025 F. Espenak, EclipseWise.com
A 1524 Feb 04
A 1504 Sep 08
T 1511 Apr 27

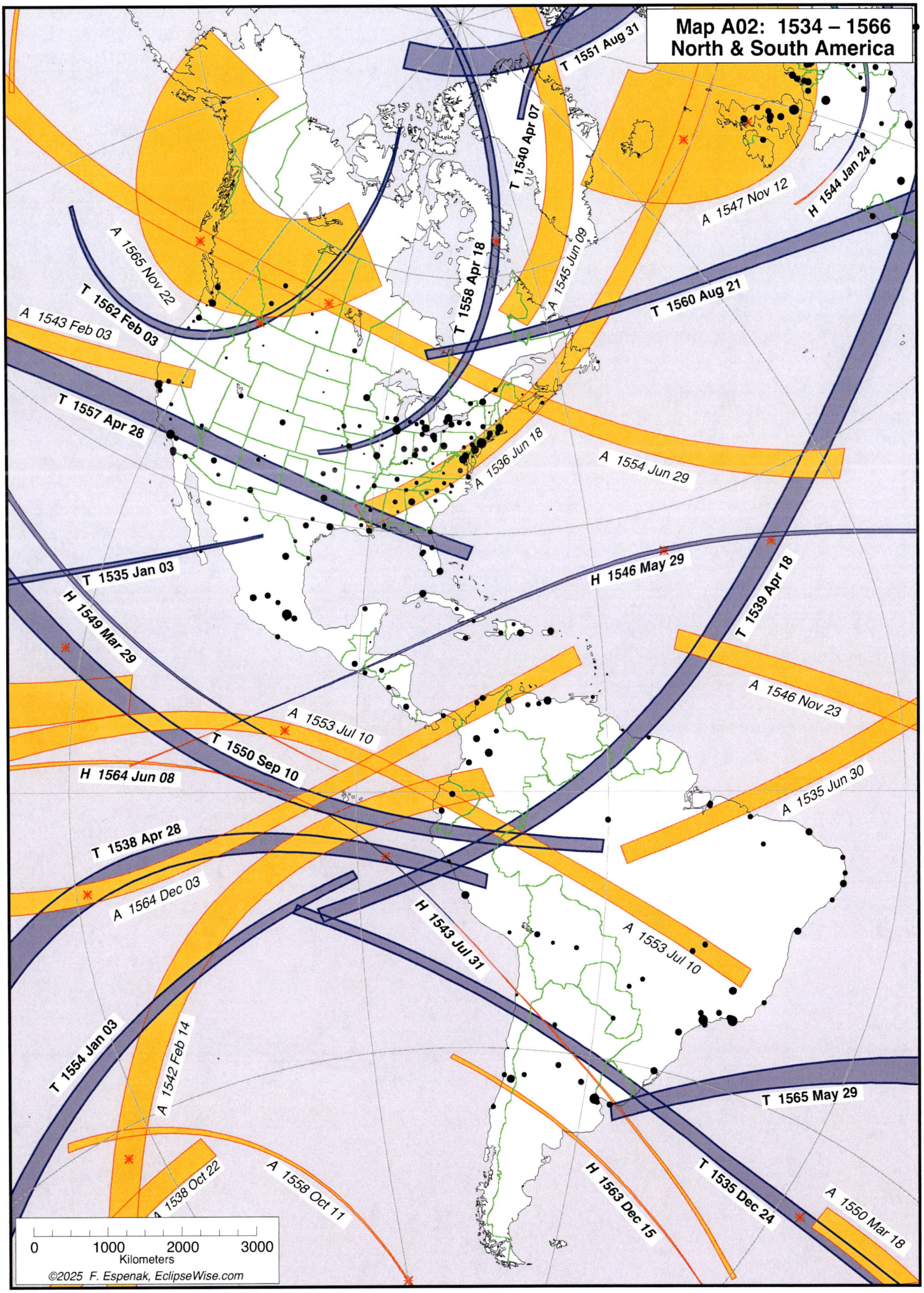
Map A02: 1534 – 1566
North & South America
T 1551 Aug 31
T 1540 Apr 07
A 1547 Nov 12
H 1544 Jan 24
A 1565 Nov 22
T 1558 Apr 18
A 1545 Jun 09
T 1562 Feb 03
A 1543 Feb 03
T 1560 Aug 21
T 1557 Apr 28
A 1536 Jun 18
A 1554 Jun 29
T 1535 Jan 03
H 1546 May 29
H 1549 Mar 29
T 1539 Apr 18
A 1546 Nov 23
A 1553 Jul 10
T 1550 Sep 10
H 1564 Jun 08
A 1535 Jun 30
T 1538 Apr 28
A 1564 Dec 03
H 1543 Jul 31
A 1553 Jul 10
T 1554 Jan 03
A 1542 Feb 14
T 1565 May 29
A 1538 Oct 22
A 1558 Oct 11
H 1563 Dec 15
T 1535 Dec 24
A 1550 Mar 18
0 1000 2000 3000
Kilometers
©2025 F. Espenak, EclipseWise.com

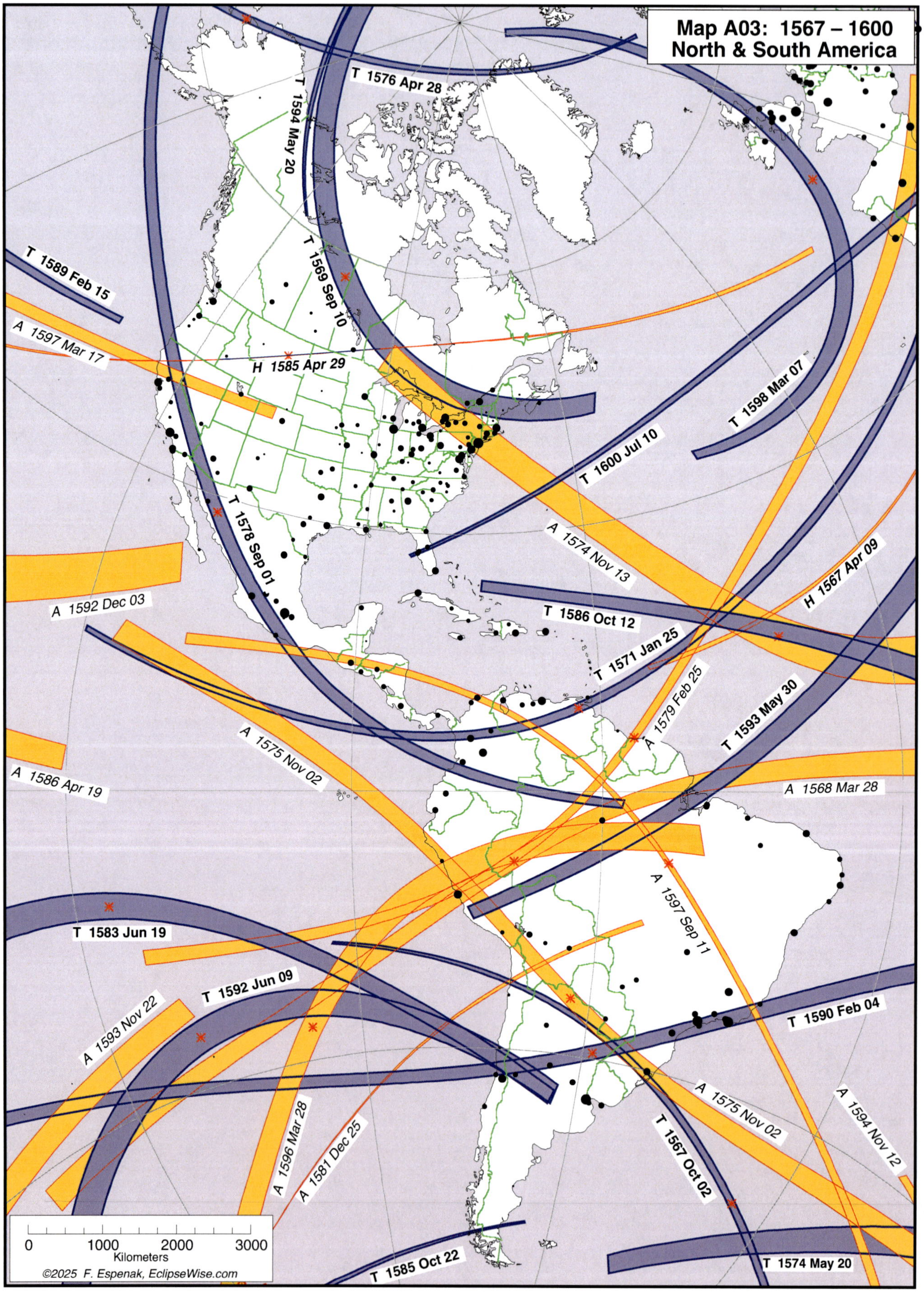
Map A03: 1567 – 1600
North & South America
T 1576 Apr 28
T 1594 May 20
T 1589 Feb 15
T 1569 Sep 10
A 1597 Mar 17
H 1585 Apr 29
T 1598 Mar 07
T 1600 Jul 10
T 1578 Sep 01
A 1574 Nov 13
A 1592 Dec 03
H 1567 Apr 09
T 1586 Oct 12
T 1571 Jan 25
A 1579 Feb 25
T 1593 May 30
A 1575 Nov 02
A 1586 Apr 19
A 1568 Mar 28
A 1597 Sep 11
T 1583 Jun 19
T 1592 Jun 09
T 1590 Feb 04
A 1593 Nov 22
A 1575 Nov 02
A 1594 Nov 12
A 1596 Mar 28
A 1581 Dec 25
T 1567 Oct 02
T 1585 Oct 22
T 1574 May 20
0 1000 2000 3000
Kilometers
©2025 F. Espenak, EclipseWise.com

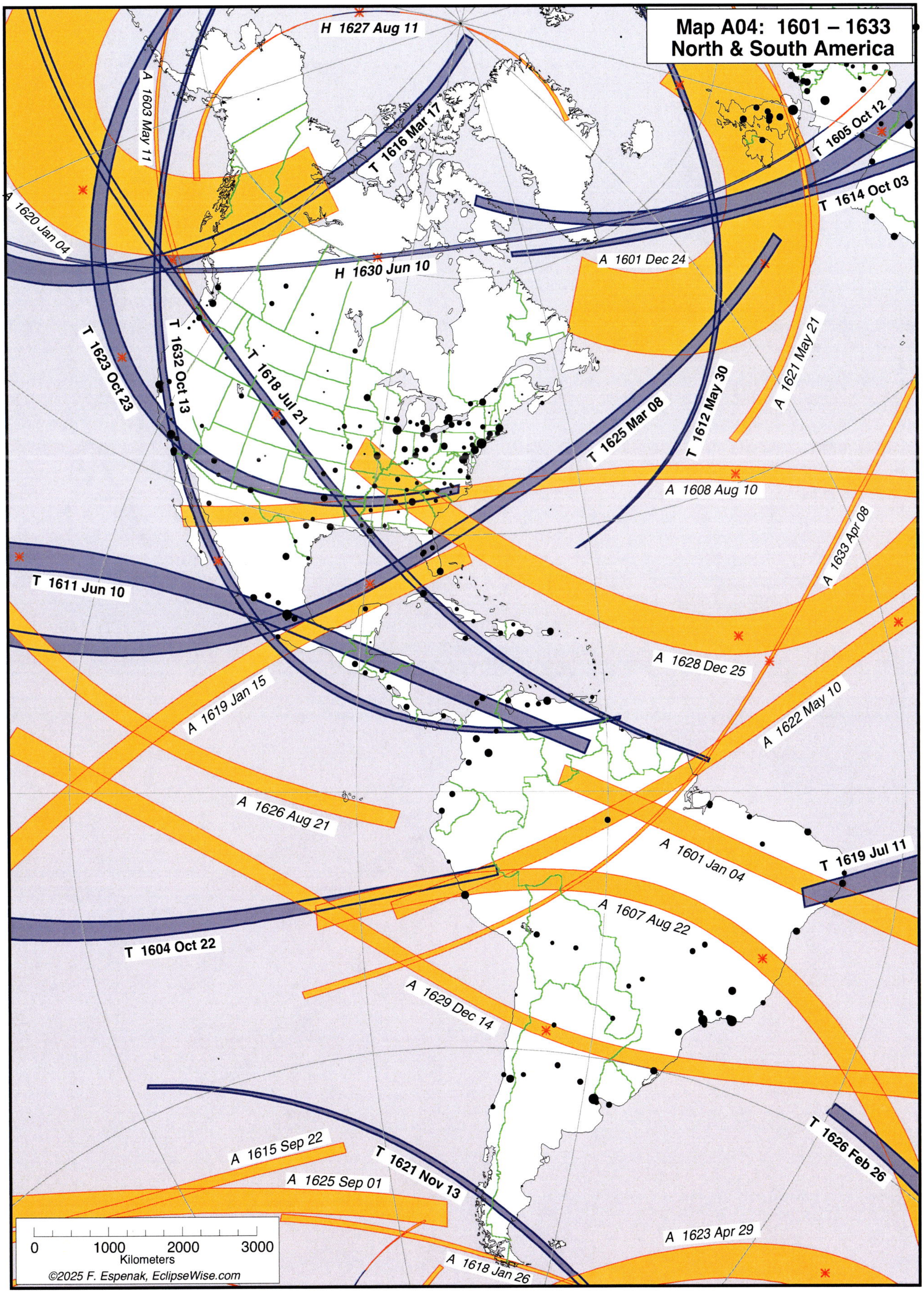
Map A04: 1601 – 1633
North & South America
H 1627 Aug 11
A 1603 May 11
T 1616 Mar 17
T 1605 Oct 12
T 1614 Oct 03
A 1620 Jan 04
A 1601 Dec 24
H 1630 Jun 10
T 1623 Oct 23
T 1632 Oct 13
T 1618 Jul 21
T 1625 Mar 08
T 1612 May 30
A 1621 May 21
A 1608 Aug 10
A 1633 Apr 08
T 1611 Jun 10
A 1628 Dec 25
A 1619 Jan 15
A 1622 May 10
A 1626 Aug 21
A 1601 Jan 04
T 1619 Jul 11
A 1607 Aug 22
T 1604 Oct 22
A 1629 Dec 14
T 1626 Feb 26
A 1615 Sep 22
T 1621 Nov 13
A 1625 Sep 01
A 1623 Apr 29
A 1618 Jan 26
0 1000 2000 3000
Kilometers
©2025 F. Espenak, EclipseWise.com

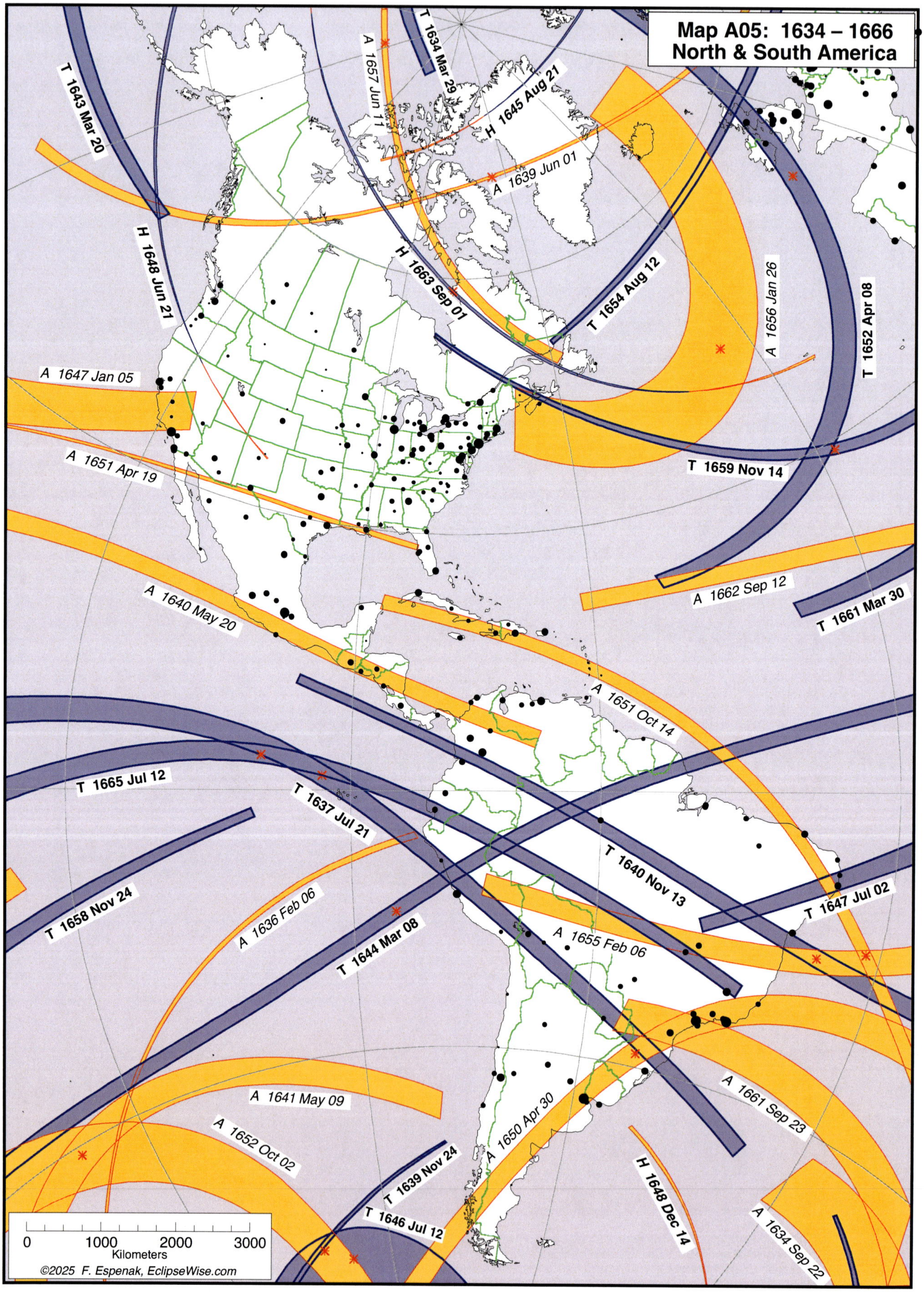
Map A05: 1634 – 1666
North & South America
T 1643 Mar 20
A 1657 Jun 11
T 1634 Mar 29
H 1645 Aug 21
A 1639 Jun 01
H 1648 Jun 21
H 1663 Sep 01
T 1654 Aug 12
A 1656 Jan 26
T 1652 Apr 08
A 1647 Jan 05
A 1651 Apr 19
T 1659 Nov 14
A 1640 May 20
A 1662 Sep 12
T 1661 Mar 30
A 1651 Oct 14
T 1665 Jul 12
T 1637 Jul 21
T 1640 Nov 13
T 1658 Nov 24
A 1636 Feb 06
T 1644 Mar 08
A 1655 Feb 06
T 1647 Jul 02
A 1641 May 09
A 1652 Oct 02
A 1650 Apr 30
A 1661 Sep 23
T 1639 Nov 24
T 1646 Jul 12
H 1648 Dec 14
A 1634 Sep 22
0
1000
2000
3000
Kilometers
©2025 F. Espenak, EclipseWise.com

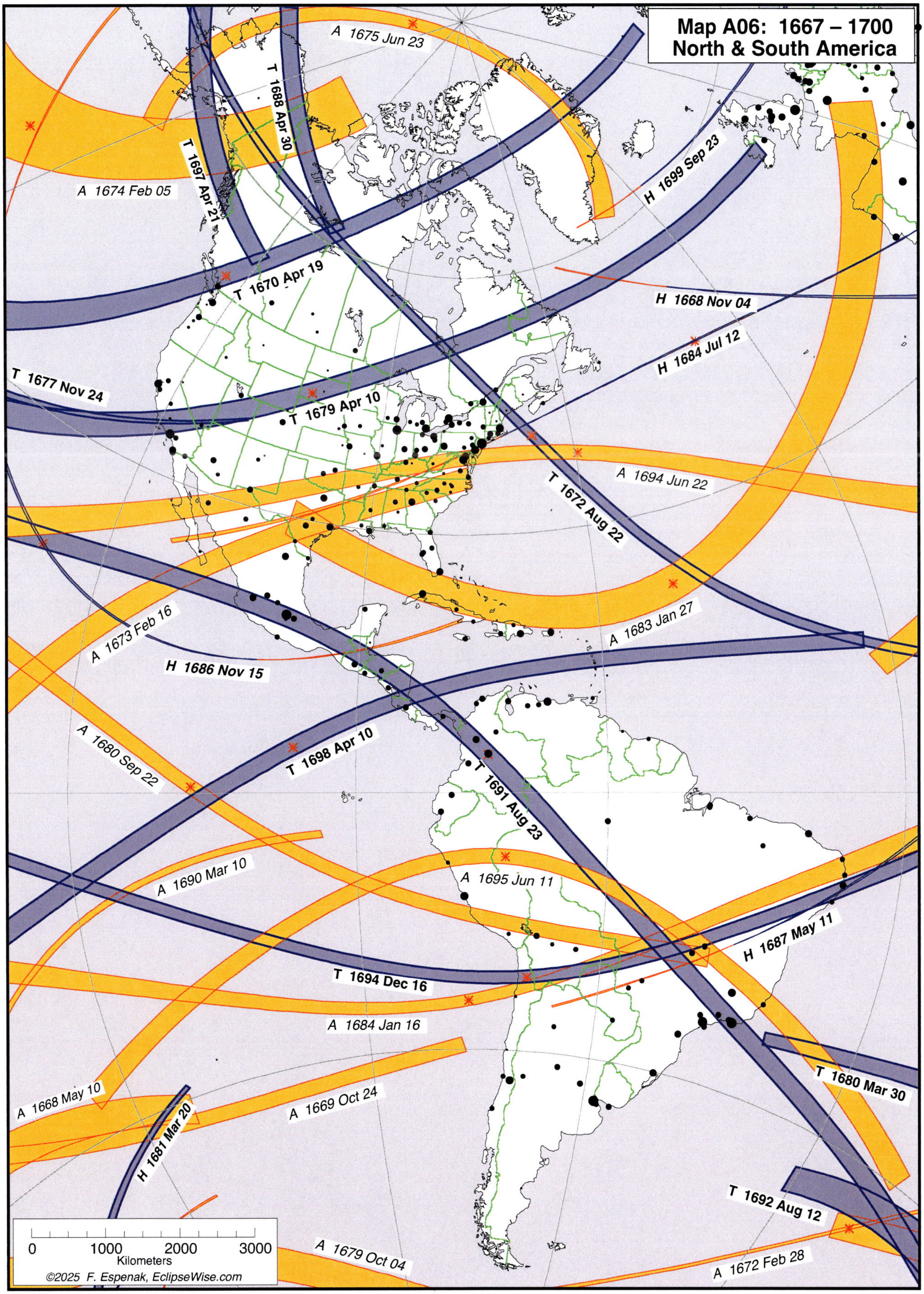
Map A06: 1667 – 1700
North & South America
A 1675 Jun 23
T 1688 Apr 30
T 1697 Apr 21
A 1674 Feb 05
H 1699 Sep 23
T 1670 Apr 19
H 1668 Nov 04
H 1684 Jul 12
T 1677 Nov 24
T 1679 Apr 10
A 1694 Jun 22
T 1672 Aug 22
A 1683 Jan 27
A 1673 Feb 16
H 1686 Nov 15
A 1680 Sep 22
T 1698 Apr 10
T 1691 Aug 23
A 1690 Mar 10
A 1695 Jun 11
H 1687 May 11
T 1694 Dec 16
A 1684 Jan 16
T 1680 Mar 30
A 1668 May 10
A 1669 Oct 24
H 1681 Mar 20
T 1692 Aug 12
A 1679 Oct 04
A 1672 Feb 28
0 1000 2000 3000
Kilometers
©2025 F. Espenak, EclipseWise.com

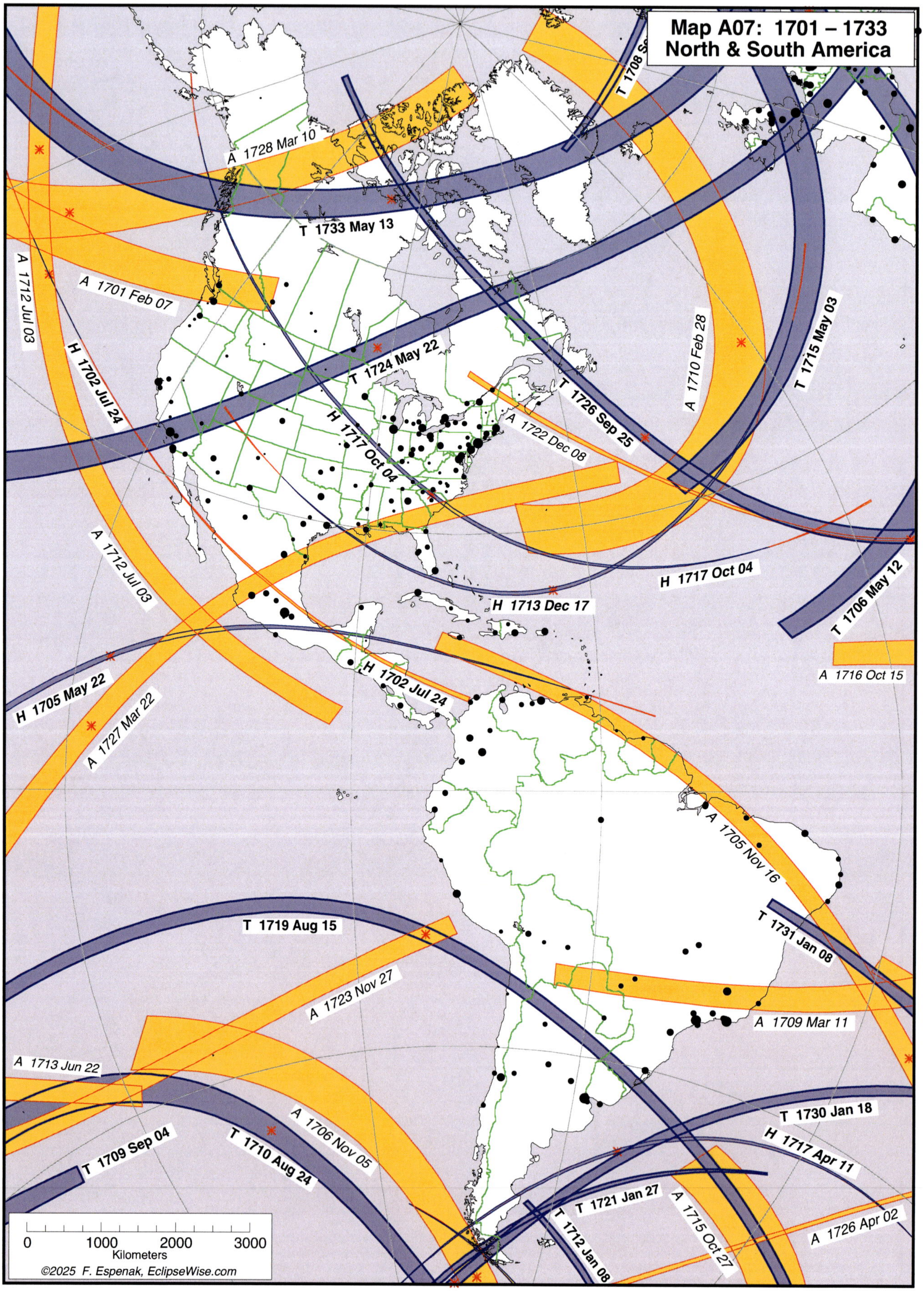
Map A07: 1701 – 1733
North & South America
A 1728 Mar 10
T 1733 May 13
A 1712 Jul 03
A 1701 Feb 07
H 1702 Jul 24
T 1724 May 22
H 1717 Oct 04
A 1722 Dec 08
T 1726 Sep 25
A 1710 Feb 28
T 1715 May 03
H 1717 Oct 04
H 1713 Dec 17
T 1706 May 12
A 1712 Jul 03
A 1716 Oct 15
H 1705 May 22
A 1727 Mar 22
H 1702 Jul 24
A 1705 Nov 16
T 1719 Aug 15
T 1731 Jan 08
A 1723 Nov 27
A 1709 Mar 11
A 1713 Jun 22
T 1709 Sep 04
T 1710 Aug 24
A 1706 Nov 05
T 1730 Jan 18
H 1717 Apr 11
T 1721 Jan 27
T 1712 Jan 08
A 1715 Oct 27
A 1726 Apr 02
0 1000 2000 3000
Kilometers
©2025 F. Espenak, EclipseWise.com

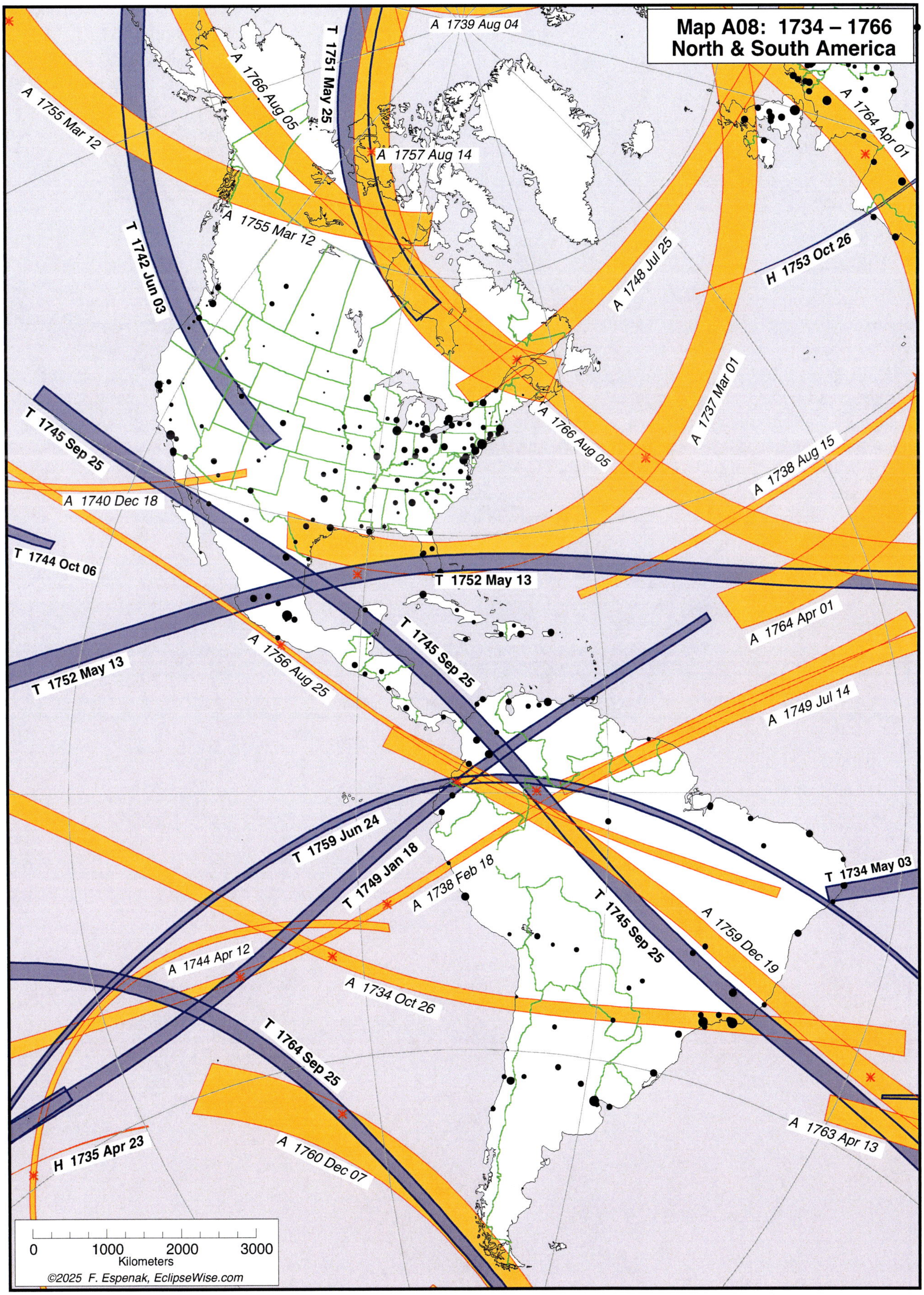
Map A08: 1734 – 1766
North & South America
A 1739 Aug 04
T 1751 May 25
A 1766 Aug 05
A 1755 Mar 12
A 1757 Aug 14
A 1764 Apr 01
A 1755 Mar 12
T 1742 Jun 03
A 1748 Jul 25
H 1753 Oct 26
A 1737 Mar 01
T 1745 Sep 25
A 1766 Aug 05
A 1738 Aug 15
A 1740 Dec 18
T 1744 Oct 06
T 1752 May 13
T 1745 Sep 25
A 1764 Apr 01
A 1756 Aug 25
T 1752 May 13
A 1749 Jul 14
T 1759 Jun 24
T 1749 Jan 18
A 1738 Feb 18
T 1734 May 03
T 1745 Sep 25
A 1759 Dec 19
A 1744 Apr 12
A 1734 Oct 26
T 1764 Sep 25
A 1760 Dec 07
A 1763 Apr 13
H 1735 Apr 23
0 1000 2000 3000
Kilometers
©2025 F. Espenak, EclipseWise.com

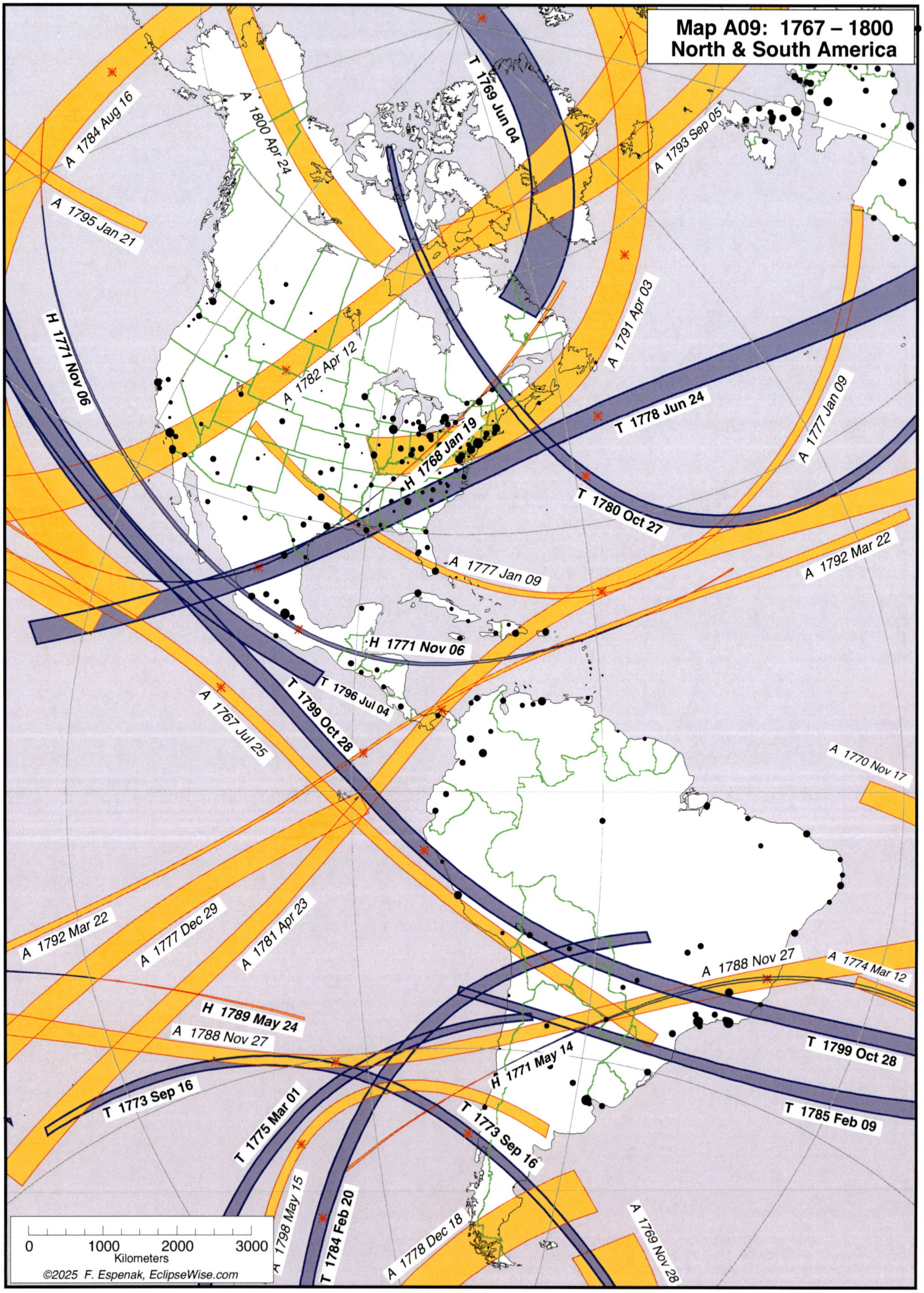
Map A09: 1767 – 1800
North & South America
A 1784 Aug 16
A 1800 Apr 24
T 1769 Jun 04
A 1793 Sep 05
A 1795 Jan 21
A 1791 Apr 03
H 1771 Nov 06
A 1782 Apr 12
T 1778 Jun 24
H 1768 Jan 19
A 1777 Jan 09
T 1780 Oct 27
A 1777 Jan 09
A 1792 Mar 22
H 1771 Nov 06
T 1796 Jul 04
A 1767 Jul 25
T 1799 Oct 28
A 1770 Nov 17
A 1792 Mar 22
A 1777 Dec 29
A 1781 Apr 23
A 1788 Nov 27
A 1774 Mar 12
H 1789 May 24
A 1788 Nov 27
T 1799 Oct 28
H 1771 May 14
T 1773 Sep 16
T 1785 Feb 09
T 1775 Mar 01
T 1773 Sep 16
A 1798 May 15
T 1784 Feb 20
A 1778 Dec 18
A 1769 Nov 28
0 1000 2000 3000
Kilometers
©2025 F. Espenak, EclipseWise.com

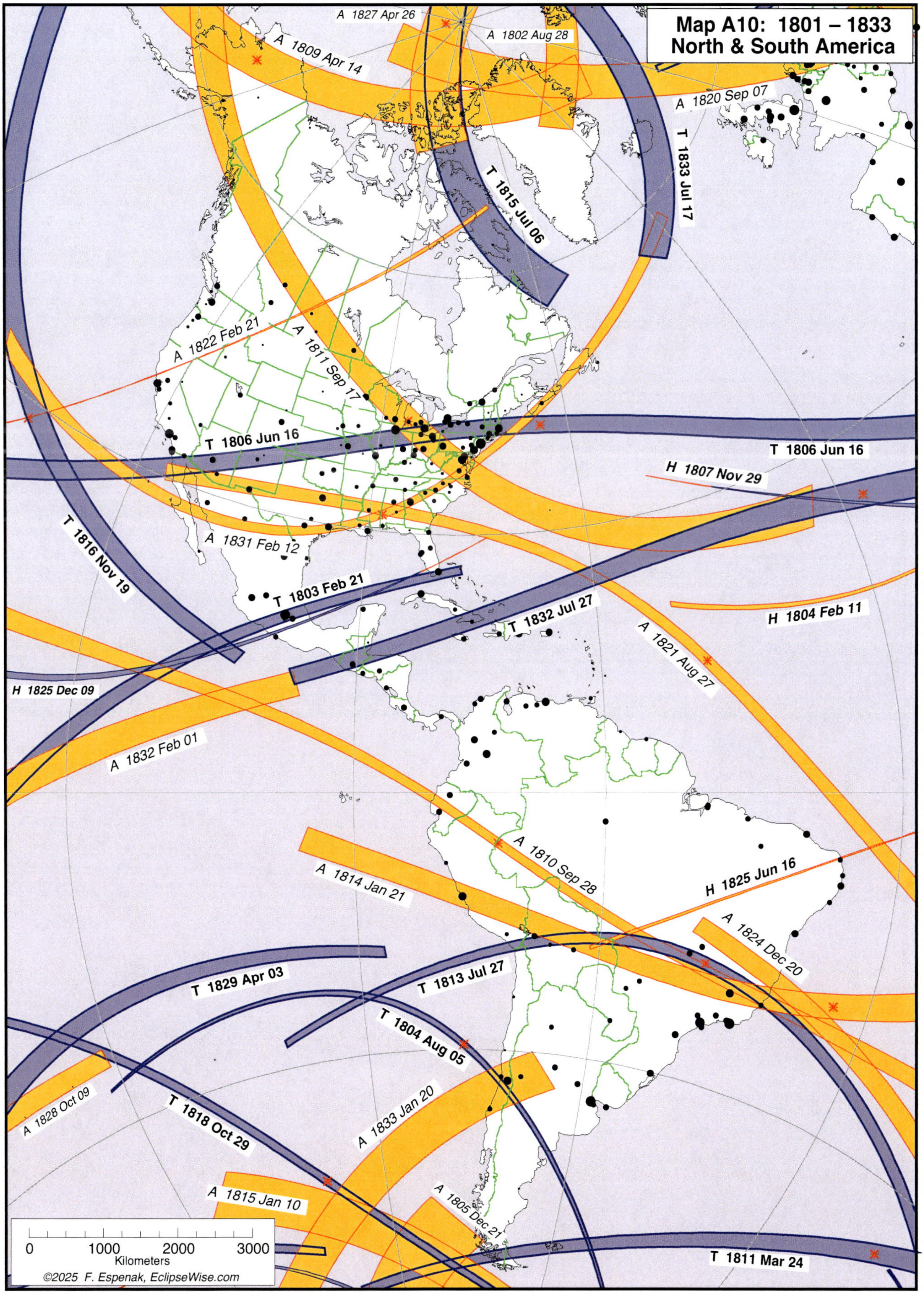
Map A10: 1801 – 1833
North & South America
A 1827 Apr 26
A 1802 Aug 28
A 1809 Apr 14
A 1820 Sep 07
T 1833 Jul 17
T 1815 Jul 06
A 1822 Feb 21
A 1811 Sep 17
T 1806 Jun 16
T 1806 Jun 16
H 1807 Nov 29
T 1816 Nov 19
A 1831 Feb 12
T 1803 Feb 21
T 1832 Jul 27
H 1804 Feb 11
A 1821 Aug 27
H 1825 Dec 09
A 1832 Feb 01
A 1810 Sep 28
A 1814 Jan 21
H 1825 Jun 16
A 1824 Dec 20
T 1829 Apr 03
T 1813 Jul 27
T 1804 Aug 05
A 1828 Oct 09
T 1818 Oct 29
A 1833 Jan 20
A 1815 Jan 10
A 1805 Dec 21
T 1811 Mar 24
0
1000
2000
3000
Kilometers
©2025 F. Espenak, EclipseWise.com

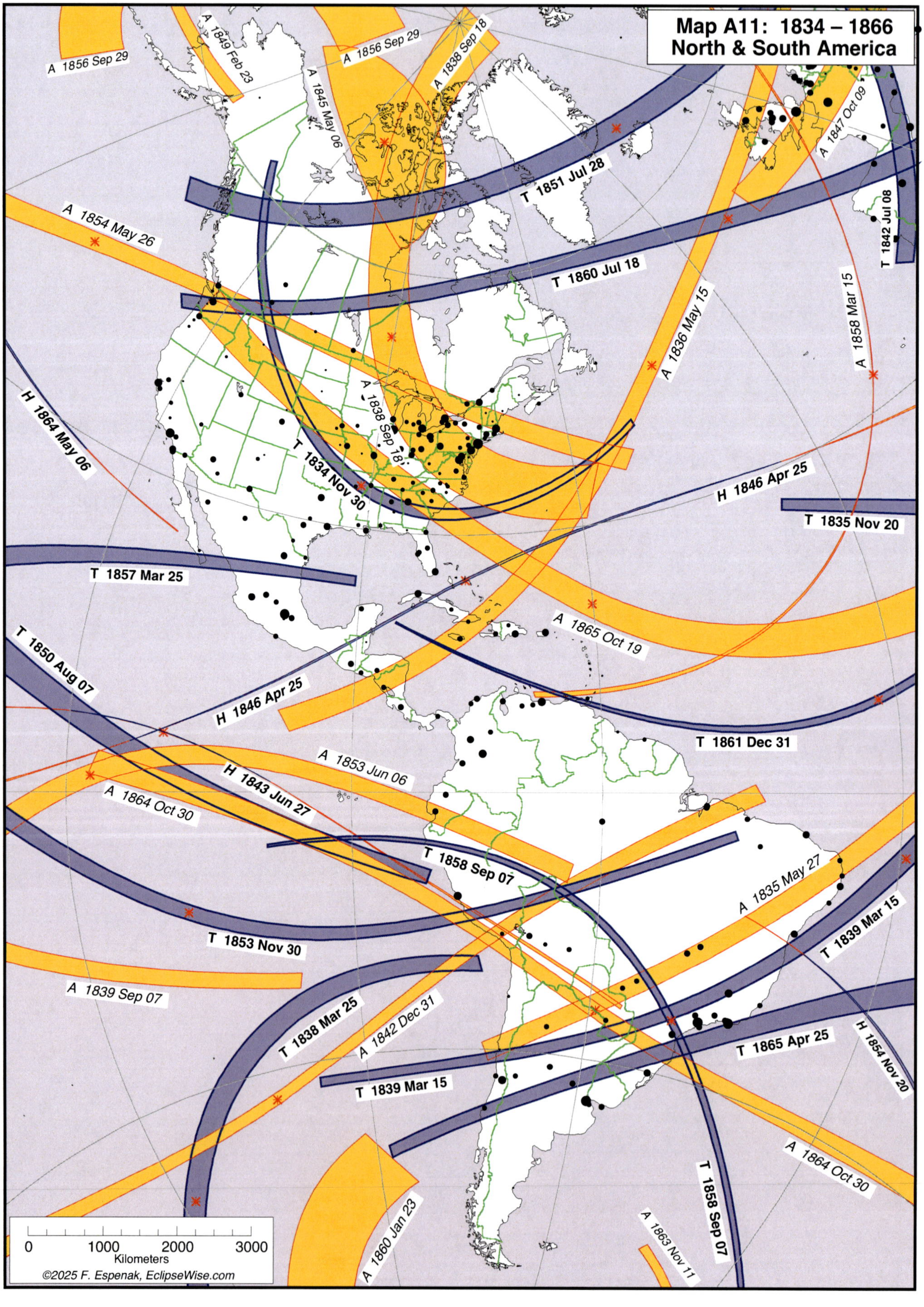

Map A11: 1834 – 1866
North & South America
A 1856 Sep 29
A 1849 Feb 23
A 1856 Sep 29
A 1838 Sep 18
A 1845 May 06
A 1847 Oct 09
T 1851 Jul 28
T 1842 Jul 08
A 1854 May 26
T 1860 Jul 18
A 1836 May 15
A 1858 Mar 15
H 1864 May 06
A 1838 Sep 18
T 1834 Nov 30
H 1846 Apr 25
T 1835 Nov 20
T 1857 Mar 25
A 1865 Oct 19
T 1850 Aug 07
H 1846 Apr 25
T 1861 Dec 31
A 1853 Jun 06
H 1843 Jun 27
A 1864 Oct 30
T 1858 Sep 07
A 1835 May 27
T 1839 Mar 15
T 1853 Nov 30
A 1839 Sep 07
T 1838 Mar 25
A 1842 Dec 31
T 1865 Apr 25
H 1854 Nov 20
T 1839 Mar 15
A 1864 Oct 30
T 1858 Sep 07
A 1860 Jan 23
A 1863 Nov 11
0 1000 2000 3000
Kilometers
©2025 F. Espenak, EclipseWise.com

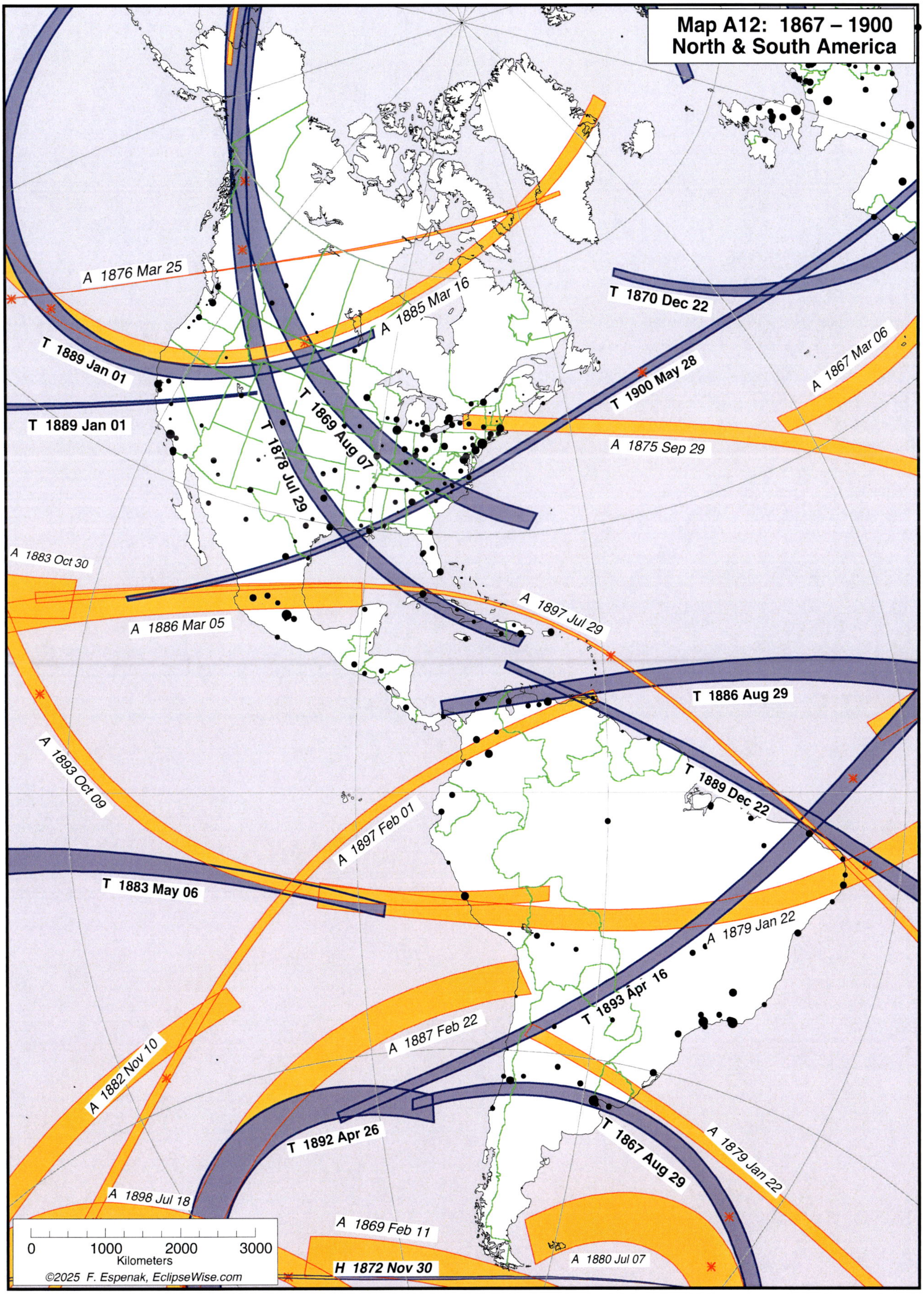

Map A12: 1867 – 1900
North & South America
A 1876 Mar 25
A 1885 Mar 16
T 1870 Dec 22
T 1889 Jan 01
T 1889 Jan 01
A 1867 Mar 06
T 1900 May 28
T 1869 Aug 07
T 1878 Jul 29
A 1875 Sep 29
A 1883 Oct 30
A 1886 Mar 05
A 1897 Jul 29
T 1886 Aug 29
A 1893 Oct 09
T 1889 Dec 22
A 1897 Feb 01
T 1883 May 06
A 1879 Jan 22
T 1893 Apr 16
A 1887 Feb 22
A 1882 Nov 10
T 1892 Apr 26
T 1867 Aug 29
A 1879 Jan 22
A 1898 Jul 18
A 1869 Feb 11
H 1872 Nov 30
A 1880 Jul 07
0 1000 2000 3000
Kilometers
©2025 F. Espenak, EclipseWise.com

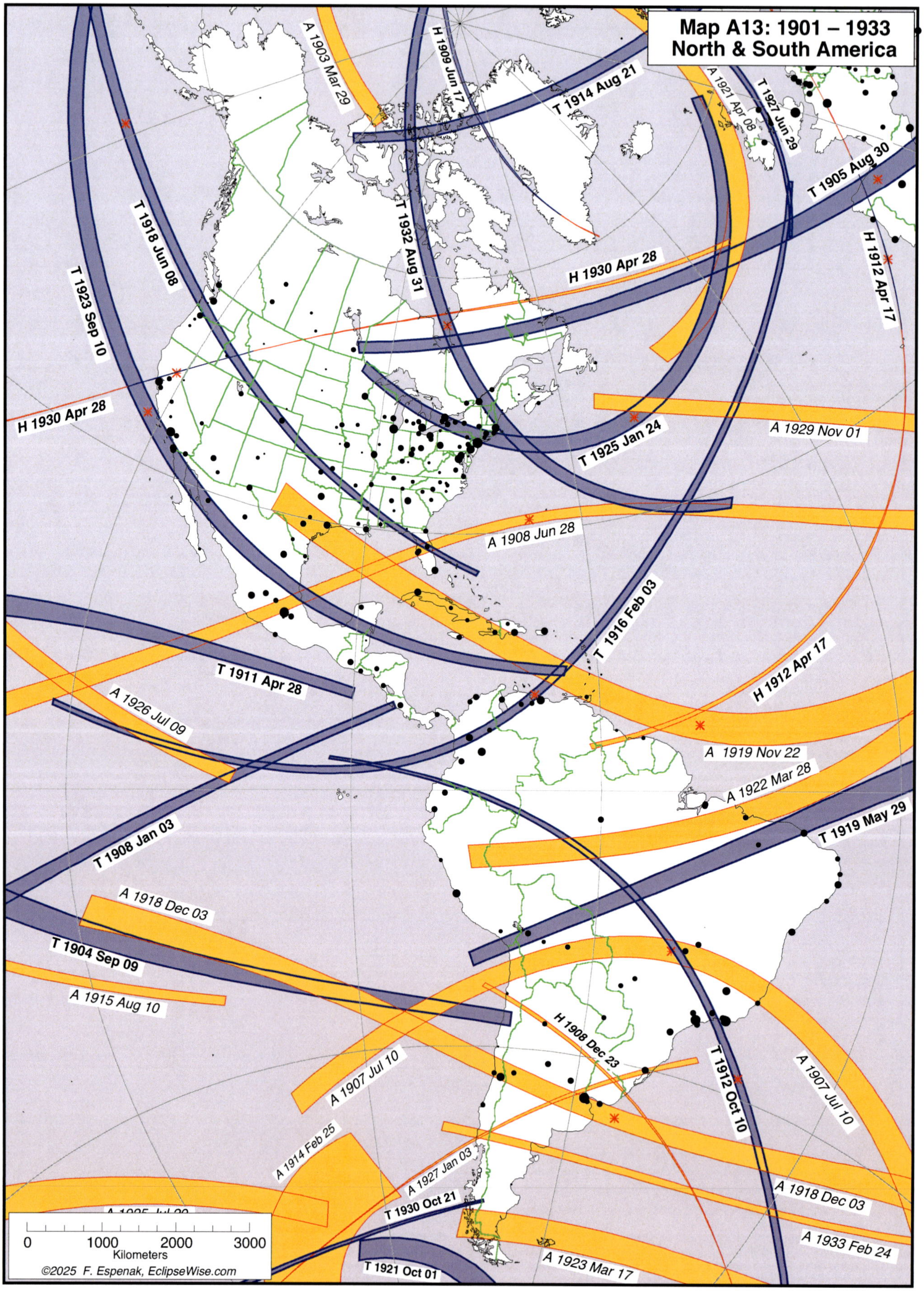
Map A13: 1901 – 1933
North & South America
A 1903 Mar 29
H 1909 Jun 17
T 1914 Aug 21
A 1921 Apr 08
T 1927 Jun 29
T 1905 Aug 30
T 1918 Jun 08
T 1932 Aug 31
H 1912 Apr 17
H 1930 Apr 28
T 1923 Sep 10
H 1930 Apr 28
A 1929 Nov 01
T 1925 Jan 24
A 1908 Jun 28
T 1916 Feb 03
H 1912 Apr 17
T 1911 Apr 28
A 1926 Jul 09
A 1919 Nov 22
A 1922 Mar 28
T 1919 May 29
T 1908 Jan 03
A 1918 Dec 03
T 1904 Sep 09
A 1915 Aug 10
H 1908 Dec 23
A 1907 Jul 10
T 1912 Oct 10
A 1907 Jul 10
A 1914 Feb 25
A 1927 Jan 03
A 1918 Dec 03
T 1930 Oct 21
A 1933 Feb 24
A 1923 Mar 17
T 1921 Oct 01
0 1000 2000 3000
Kilometers
©2025 F. Espenak, EclipseWise.com

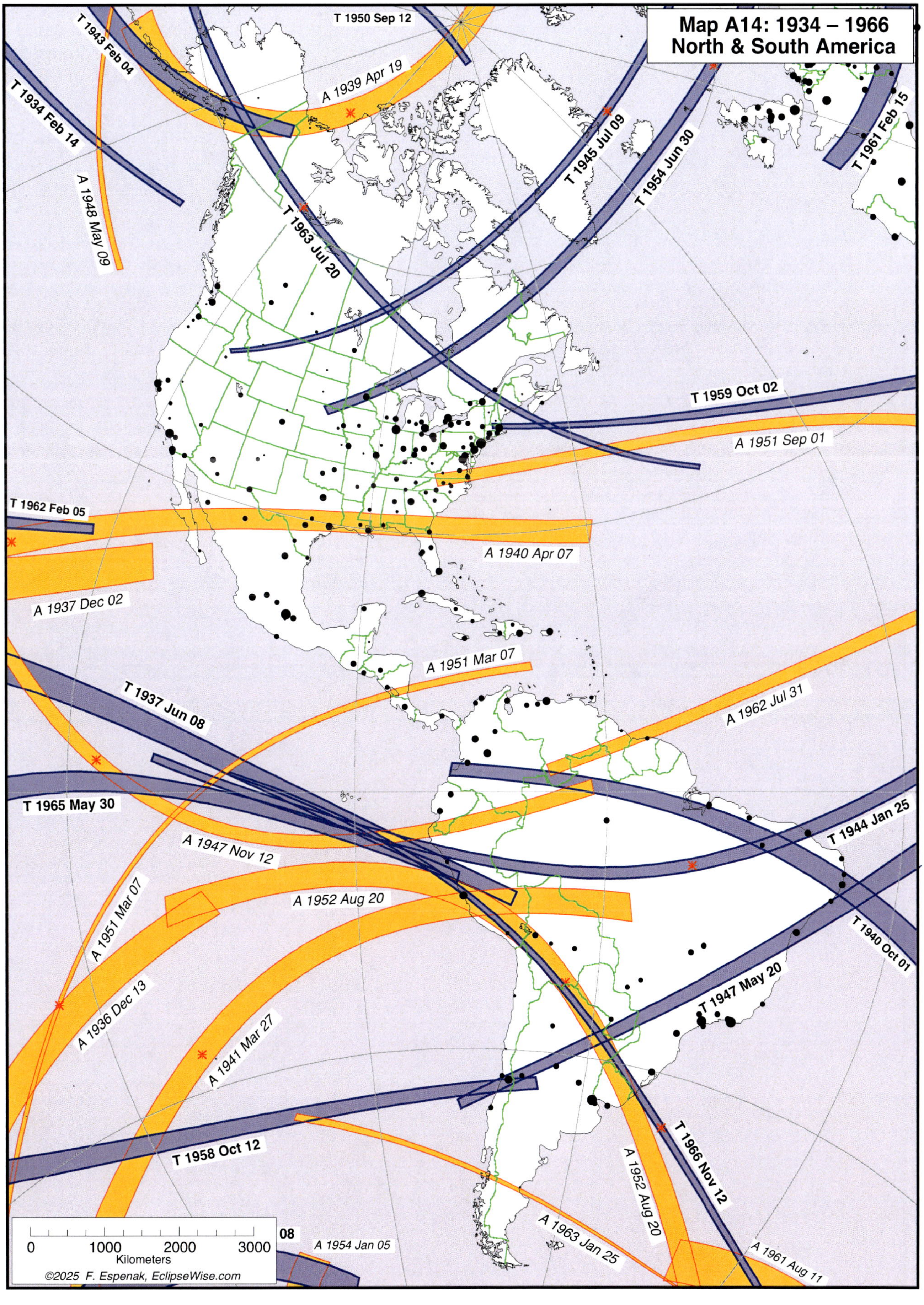
Map A14: 1934 – 1966
North & South America
T 1950 Sep 12
T 1943 Feb 04
A 1939 Apr 19
T 1934 Feb 14
A 1948 May 09
T 1945 Jul 09
T 1954 Jun 30
T 1961 Feb 15
T 1963 Jul 20
T 1959 Oct 02
A 1951 Sep 01
T 1962 Feb 05
A 1940 Apr 07
A 1937 Dec 02
A 1951 Mar 07
T 1937 Jun 08
A 1962 Jul 31
T 1965 May 30
T 1944 Jan 25
A 1947 Nov 12
A 1952 Aug 20
A 1951 Mar 07
T 1940 Oct 01
A 1936 Dec 13
T 1947 May 20
A 1941 Mar 27
T 1958 Oct 12
T 1966 Nov 12
A 1952 Aug 20
A 1963 Jan 25
08
A 1954 Jan 05
A 1961 Aug 11
0
1000
2000
3000
Kilometers
©2025 F. Espenak, EclipseWise.com

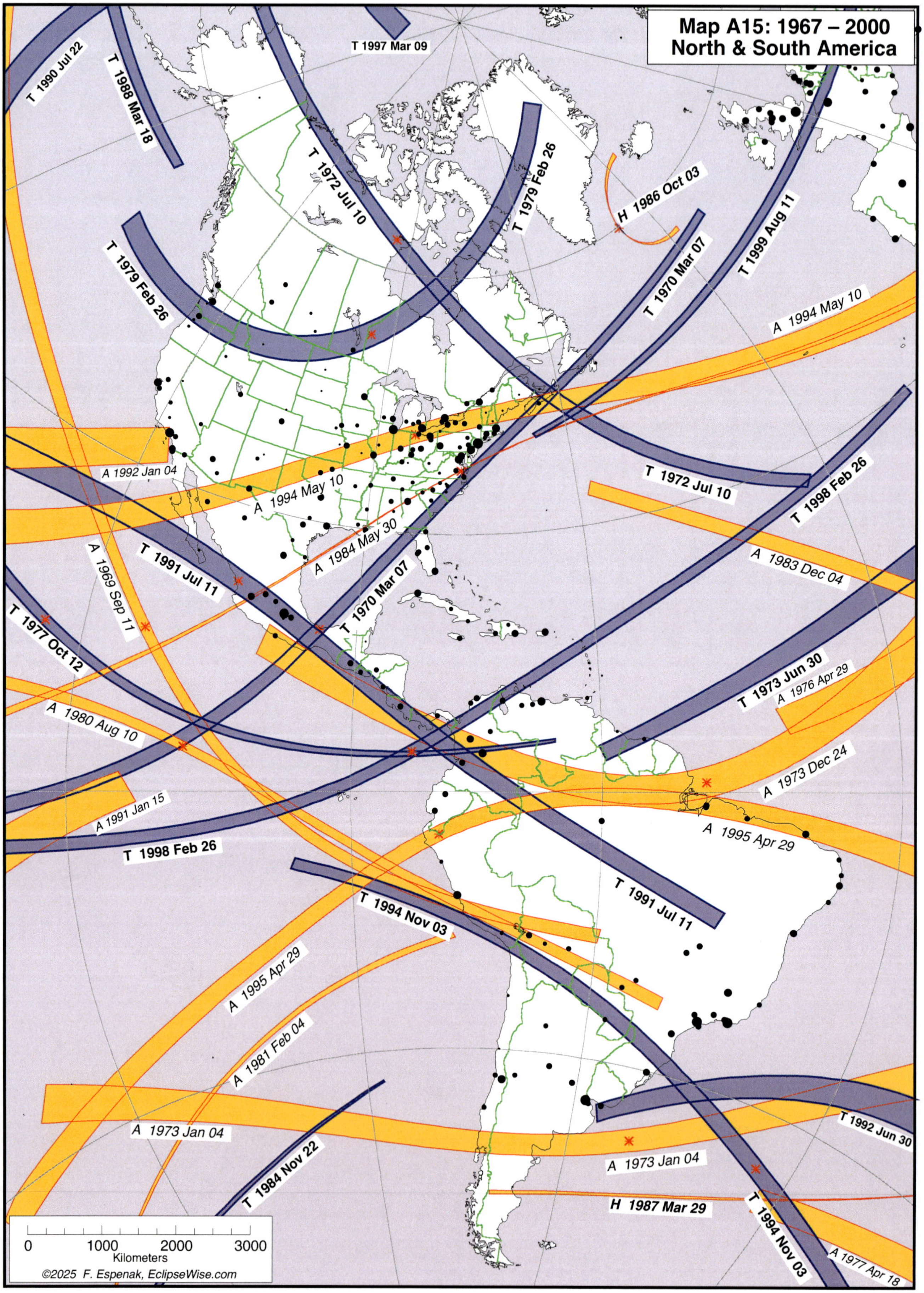
Map A15: 1967 – 2000
North & South America
T 1997 Mar 09
T 1990 Jul 22
T 1988 Mar 18
T 1972 Jul 10
T 1979 Feb 26
H 1986 Oct 03
T 1999 Aug 11
T 1970 Mar 07
T 1979 Feb 26
A 1994 May 10
A 1992 Jan 04
A 1994 May 10
T 1972 Jul 10
T 1998 Feb 26
A 1984 May 30
A 1983 Dec 04
A 1969 Sep 17
T 1991 Jul 11
T 1970 Mar 07
T 1977 Oct 12
T 1973 Jun 30
A 1976 Apr 29
A 1980 Aug 10
A 1973 Dec 24
A 1991 Jan 15
T 1998 Feb 26
A 1995 Apr 29
T 1991 Jul 11
T 1994 Nov 03
A 1995 Apr 29
A 1981 Feb 04
A 1973 Jan 04
T 1992 Jun 30
T 1984 Nov 22
A 1973 Jan 04
H 1987 Mar 29
T 1994 Nov 03
A 1977 Apr 18
0
1000
2000
3000
Kilometers
©2025 F. Espenak, EclipseWise.com

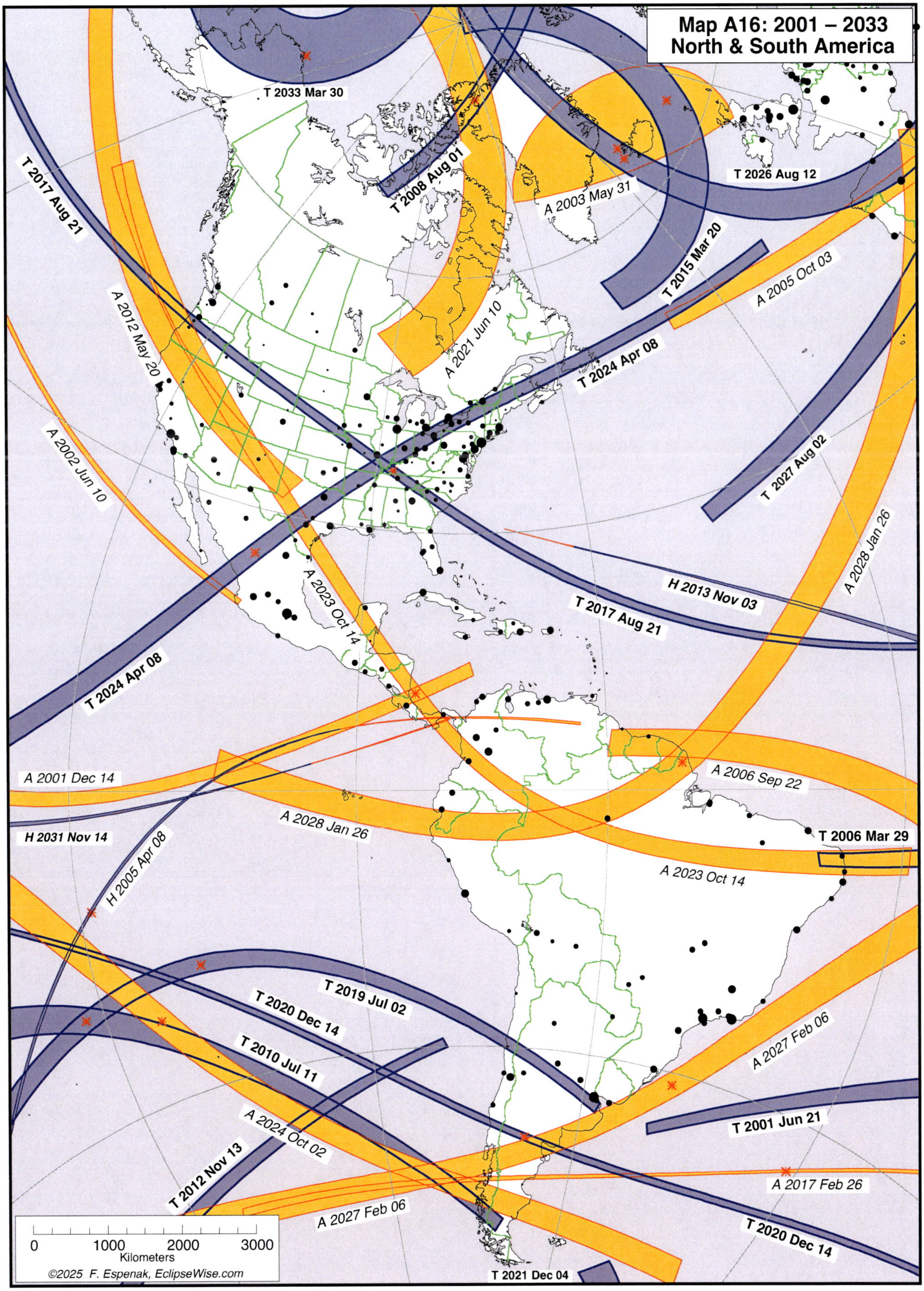
Map A16: 2001 – 2033
North & South America
T 2033 Mar 30
T 2008 Aug 01
A 2003 May 31
T 2026 Aug 12
T 2017 Aug 21
T 2015 Mar 20
A 2005 Oct 03
A 2012 May 20
A 2021 Jun 10
T 2024 Apr 08
A 2002 Jun 10
T 2027 Aug 02
A 2028 Jan 26
H 2013 Nov 03
A 2023 Oct 14
T 2017 Aug 21
T 2024 Apr 08
A 2001 Dec 14
A 2006 Sep 22
H 2031 Nov 14
A 2028 Jan 26
T 2006 Mar 29
H 2005 Apr 08
A 2023 Oct 14
T 2019 Jul 02
T 2020 Dec 14
A 2027 Feb 06
T 2010 Jul 11
T 2001 Jun 21
A 2024 Oct 02
T 2012 Nov 13
A 2017 Feb 26
A 2027 Feb 06
T 2020 Dec 14
T 2021 Dec 04
0 1000 2000 3000
Kilometers
©2025 F. Espenak, EclipseWise.com

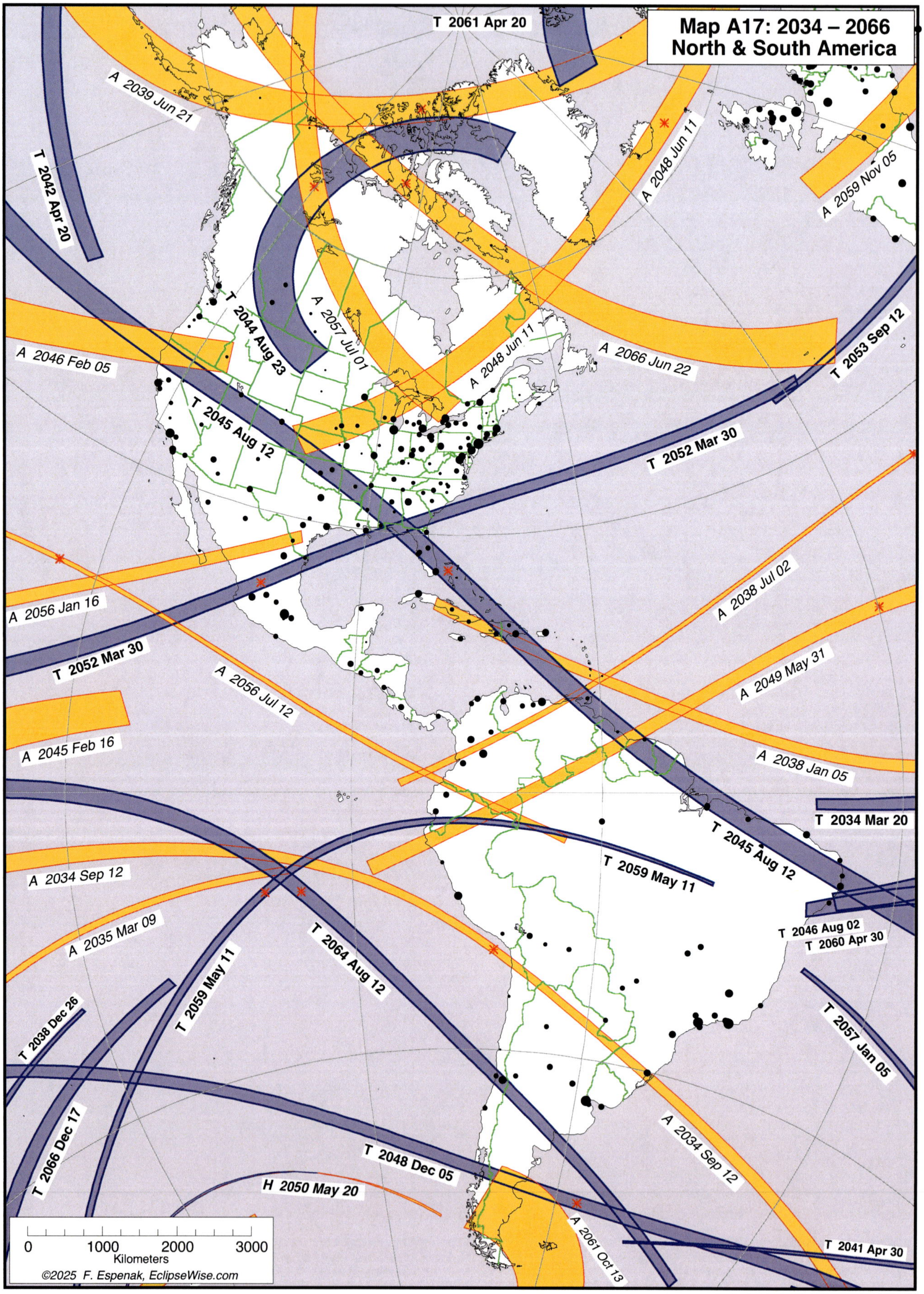
Map A17: 2034 – 2066
North & South America
T 2061 Apr 20
A 2039 Jun 21
T 2042 Apr 20
A 2048 Jun 11
A 2059 Nov 05
T 2044 Aug 23
A 2057 Jul 01
A 2046 Feb 05
A 2066 Jun 22
T 2053 Sep 12
T 2045 Aug 12
T 2052 Mar 30
A 2056 Jan 16
A 2038 Jul 02
A 2049 May 31
A 2056 Jul 12
A 2045 Feb 16
A 2038 Jan 05
T 2034 Mar 20
T 2059 May 11
A 2034 Sep 12
A 2035 Mar 09
T 2064 Aug 12
T 2046 Aug 02
T 2060 Apr 30
T 2057 Jan 05
T 2038 Dec 26
T 2066 Dec 17
T 2048 Dec 05
H 2050 May 20
A 2061 Oct 13
T 2041 Apr 30
0
1000
2000
3000
Kilometers
©2025 F. Espenak, EclipseWise.com

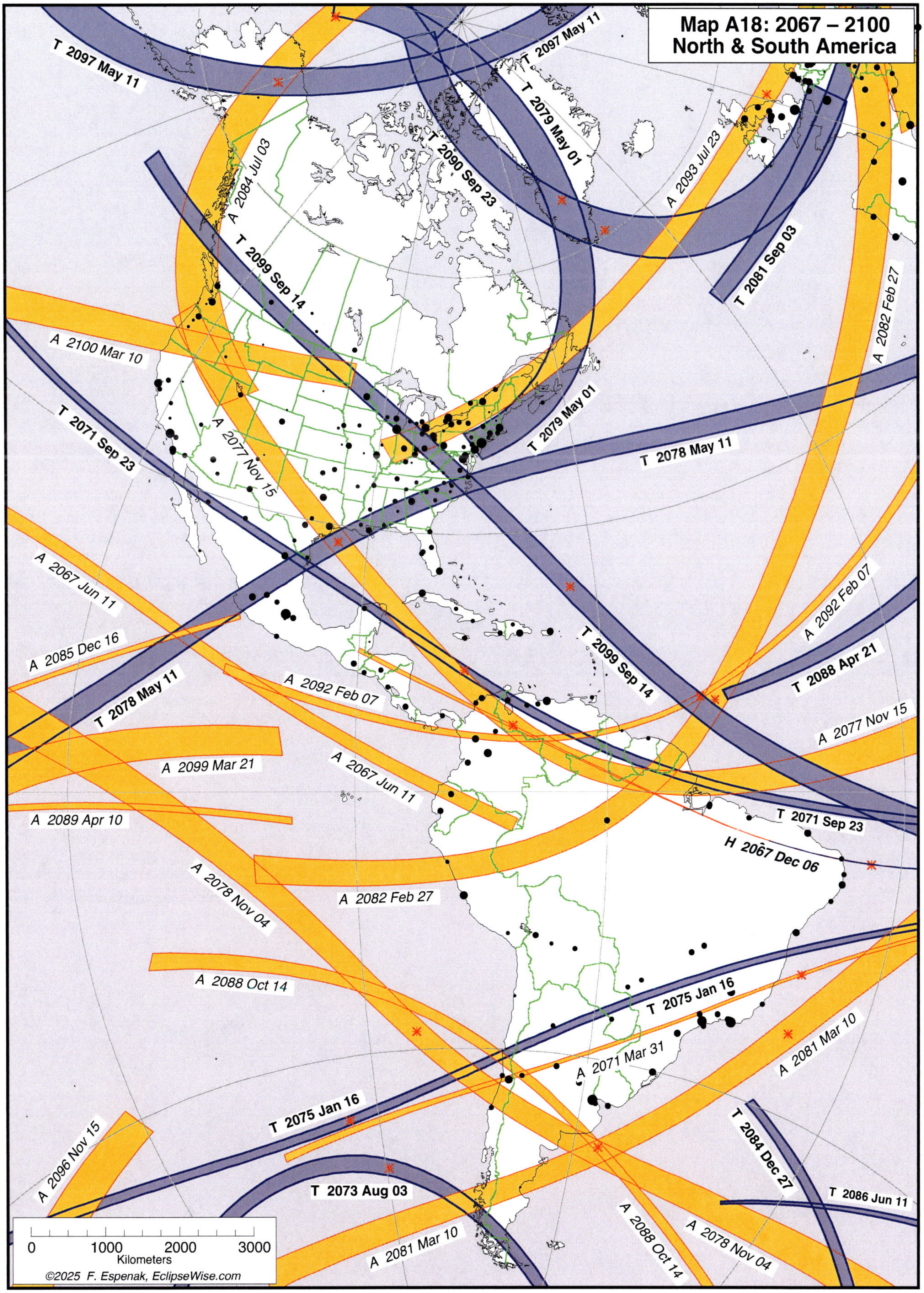
Map A18: 2067 – 2100
North & South America
T 2097 May 11
T 2097 May 11
T 2079 May 01
T 2090 Sep 23
A 2084 Jul 03
A 2093 Jul 23
T 2099 Sep 14
T 2081 Sep 03
A 2082 Feb 27
A 2100 Mar 10
T 2079 May 01
T 2071 Sep 23
A 2077 Nov 15
T 2078 May 11
A 2067 Jun 11
A 2092 Feb 07
A 2085 Dec 16
T 2099 Sep 14
T 2088 Apr 21
A 2092 Feb 07
T 2078 May 11
A 2077 Nov 15
A 2099 Mar 21
A 2067 Jun 11
A 2089 Apr 10
T 2071 Sep 23
H 2067 Dec 06
A 2078 Nov 04
A 2082 Feb 27
A 2088 Oct 14
T 2075 Jan 16
A 2071 Mar 31
A 2081 Mar 10
T 2075 Jan 16
A 2096 Nov 15
T 2084 Dec 27
T 2073 Aug 03
T 2086 Jun 11
A 2088 Oct 14
A 2078 Nov 04
A 2081 Mar 10
0
1000
2000
3000
Kilometers
©2025 F. Espenak, EclipseWise.com

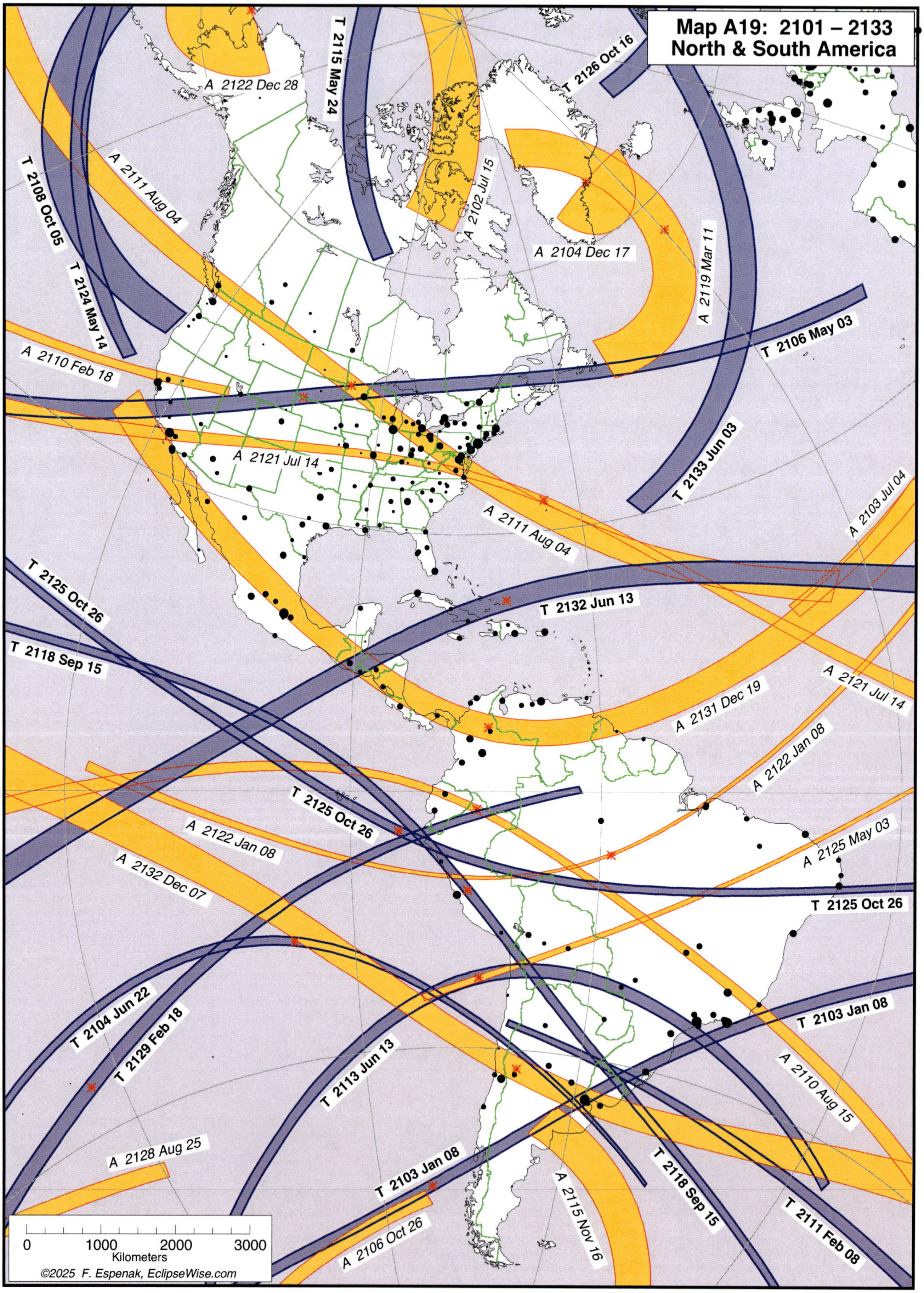
Map A19: 2101 – 2133
North & South America
A 2122 Dec 28
T 2115 May 24
T 2126 Oct 16
A 2102 Jul 15
A 2104 Dec 17
A 2119 Mar 11
T 2108 Oct 05
A 2111 Aug 04
T 2124 May 14
T 2106 May 03
A 2110 Feb 18
A 2121 Jul 14
T 2133 Jun 03
A 2111 Aug 04
A 2103 Jul 04
T 2125 Oct 26
T 2132 Jun 13
T 2118 Sep 15
A 2121 Jul 14
A 2131 Dec 19
A 2122 Jan 08
T 2125 Oct 26
A 2122 Jan 08
A 2125 May 03
A 2132 Dec 07
T 2125 Oct 26
T 2103 Jan 08
T 2104 Jun 22
T 2129 Feb 18
T 2113 Jun 13
A 2110 Aug 15
A 2128 Aug 25
T 2103 Jan 08
T 2118 Sep 15
A 2115 Nov 16
T 2111 Feb 08
A 2106 Oct 26
0 1000 2000 3000
Kilometers
©2025 F. Espenak, EclipseWise.com

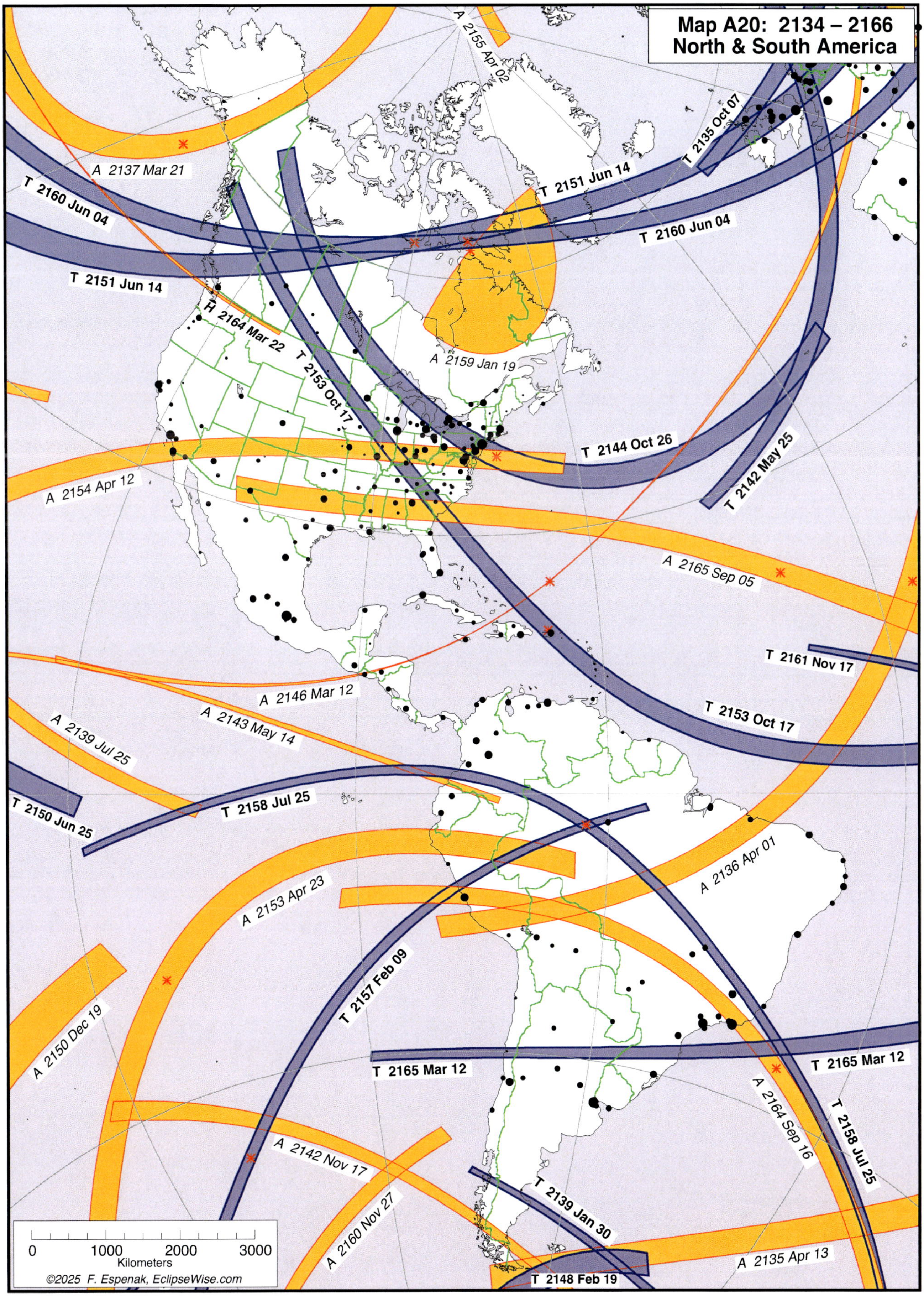
Map A20: 2134 – 2166
North & South America
A 2155 Apr 02
T 2135 Oct 07
A 2137 Mar 21
T 2160 Jun 04
T 2151 Jun 14
T 2160 Jun 04
T 2151 Jun 14
H 2164 Mar 22
T 2153 Oct 17
A 2159 Jan 19
T 2144 Oct 26
T 2142 May 25
A 2154 Apr 12
A 2165 Sep 05
T 2161 Nov 17
A 2146 Mar 12
A 2143 May 14
A 2139 Jul 25
T 2153 Oct 17
T 2158 Jul 25
T 2150 Jun 25
A 2136 Apr 01
A 2153 Apr 23
T 2157 Feb 09
A 2150 Dec 19
T 2165 Mar 12
T 2165 Mar 12
A 2164 Sep 16
T 2158 Jul 25
A 2142 Nov 17
A 2160 Nov 27
T 2139 Jan 30
A 2135 Apr 13
T 2148 Feb 19
0 1000 2000 3000
Kilometers
©2025 F. Espenak, EclipseWise.com

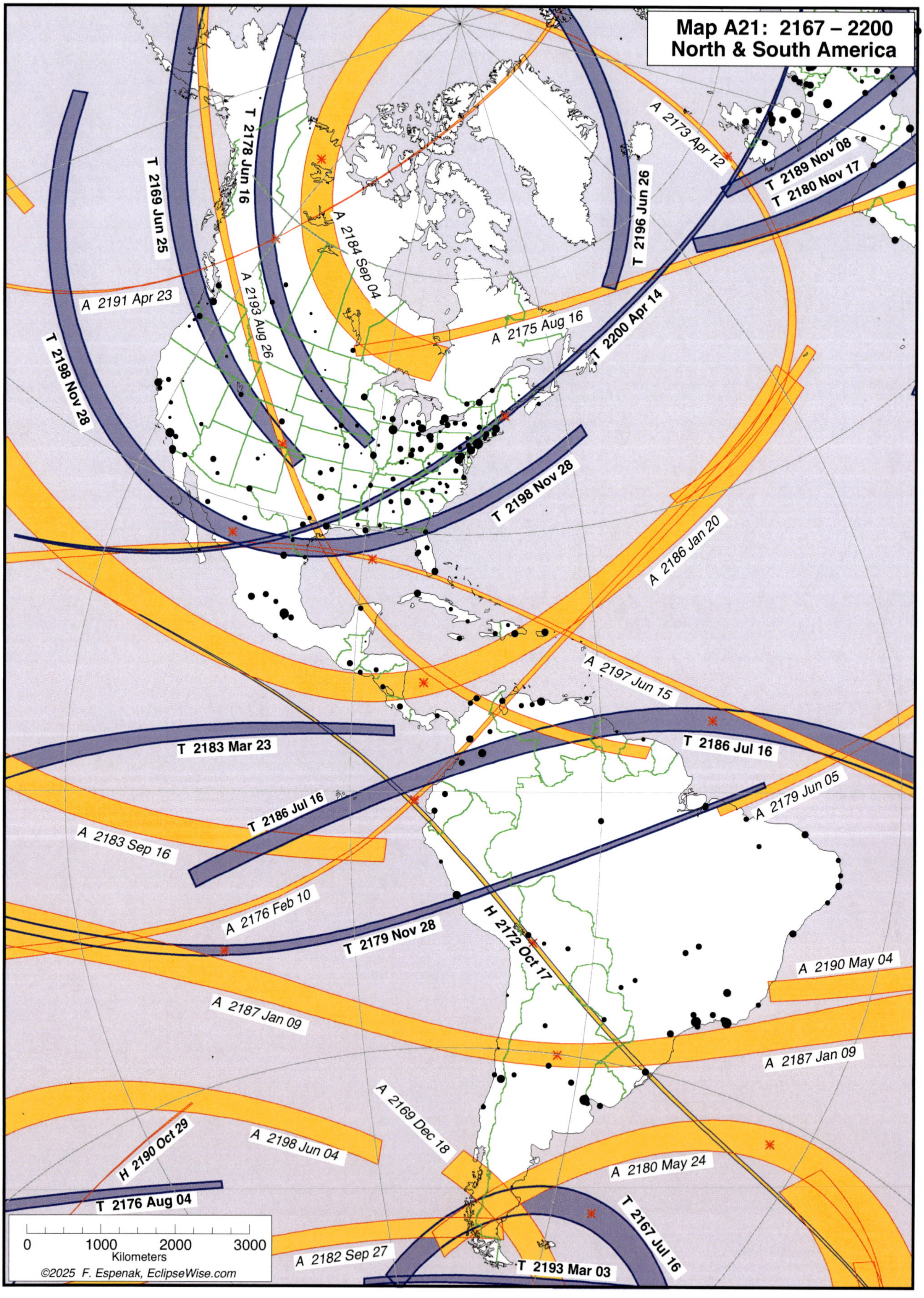
Map A21: 2167 – 2200
North & South America
T 2169 Jun 25
T 2178 Jun 16
A 2184 Sep 04
T 2196 Jun 26
A 2173 Apr 12
T 2189 Nov 08
T 2180 Nov 17
A 2191 Apr 23
A 2193 Aug 26
A 2175 Aug 16
T 2200 Apr 14
T 2198 Nov 28
T 2198 Nov 28
A 2186 Jan 20
A 2197 Jun 15
T 2183 Mar 23
T 2186 Jul 16
T 2186 Jul 16
A 2179 Jun 05
A 2183 Sep 16
A 2176 Feb 10
T 2179 Nov 28
H 2172 Oct 17
A 2190 May 04
A 2187 Jan 09
A 2187 Jan 09
A 2169 Dec 18
H 2190 Oct 29
A 2198 Jun 04
A 2180 May 24
T 2176 Aug 04
T 2167 Jul 16
0 1000 2000 3000
Kilometers
©2025 F. Espenak, EclipseWise.com
A 2182 Sep 27
T 2193 Mar 03

Map A22: 2201 – 2233
North & South America
T 2214 Jul 08
H 2209 May 05
A 2229 Sep 1
A 2204 Feb 02
A 2220 Sep 27
T 2218 Apr 25
T 2207 Nov 20
T 2232 Jul 18
A 2213 Feb 21
T 2205 Jul 17
T 2205 Jul 17
A 2230 Mar 15
A 2208 May 15
T 2223 Jul 28
A 2219 Oct 09
T 2207 Nov 20
T 2212 Aug 27
T 2219 Apr 15
T 2233 Dec 31
A 2215 Jan 01
A 2208 May 15
T 2211 Mar 15
H 2226 Nov 20
A 2212 Mar 04
T 2215 Dec 21
T 2219 Apr 15
T 2204 Jul 27
A 2207 May 27
T 2230 Sep 07
A 2205 Jan 20
A 2218 Oct 20
T 2202 Mar 24
T 2231 Aug 28
A 2209 Oct 29
0
1000
2000
3000
Kilometers
©2025 F. Espenak, EclipseWise.com

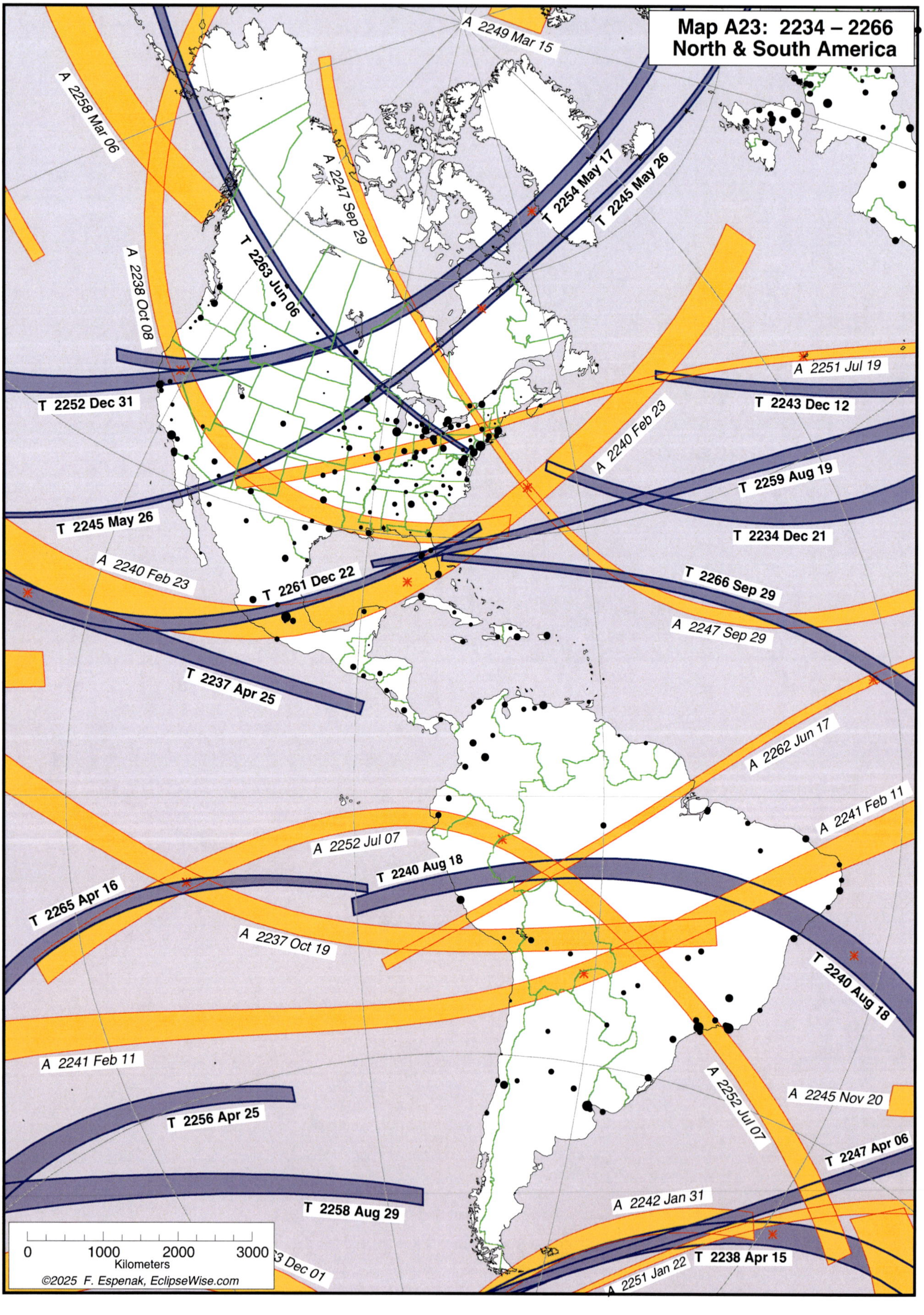
Map A23: 2234 – 2266
North & South America
A 2249 Mar 15
A 2258 Mar 06
A 2247 Sep 29
T 2254 May 17
T 2245 May 26
A 2238 Oct 08
T 2263 Jun 06
A 2251 Jul 19
T 2252 Dec 31
T 2243 Dec 12
A 2240 Feb 23
T 2259 Aug 19
T 2245 May 26
T 2234 Dec 21
A 2240 Feb 23
T 2261 Dec 22
T 2266 Sep 29
A 2247 Sep 29
T 2237 Apr 25
A 2262 Jun 17
A 2241 Feb 11
A 2252 Jul 07
T 2240 Aug 18
T 2265 Apr 16
A 2237 Oct 19
T 2240 Aug 18
A 2241 Feb 11
A 2252 Jul 07
A 2245 Nov 20
T 2256 Apr 25
T 2247 Apr 06
A 2242 Jan 31
T 2258 Aug 29
0 1000 2000 3000
Kilometers
©2025 F. Espenak, EclipseWise.com
A 2251 Jan 22
T 2238 Apr 15

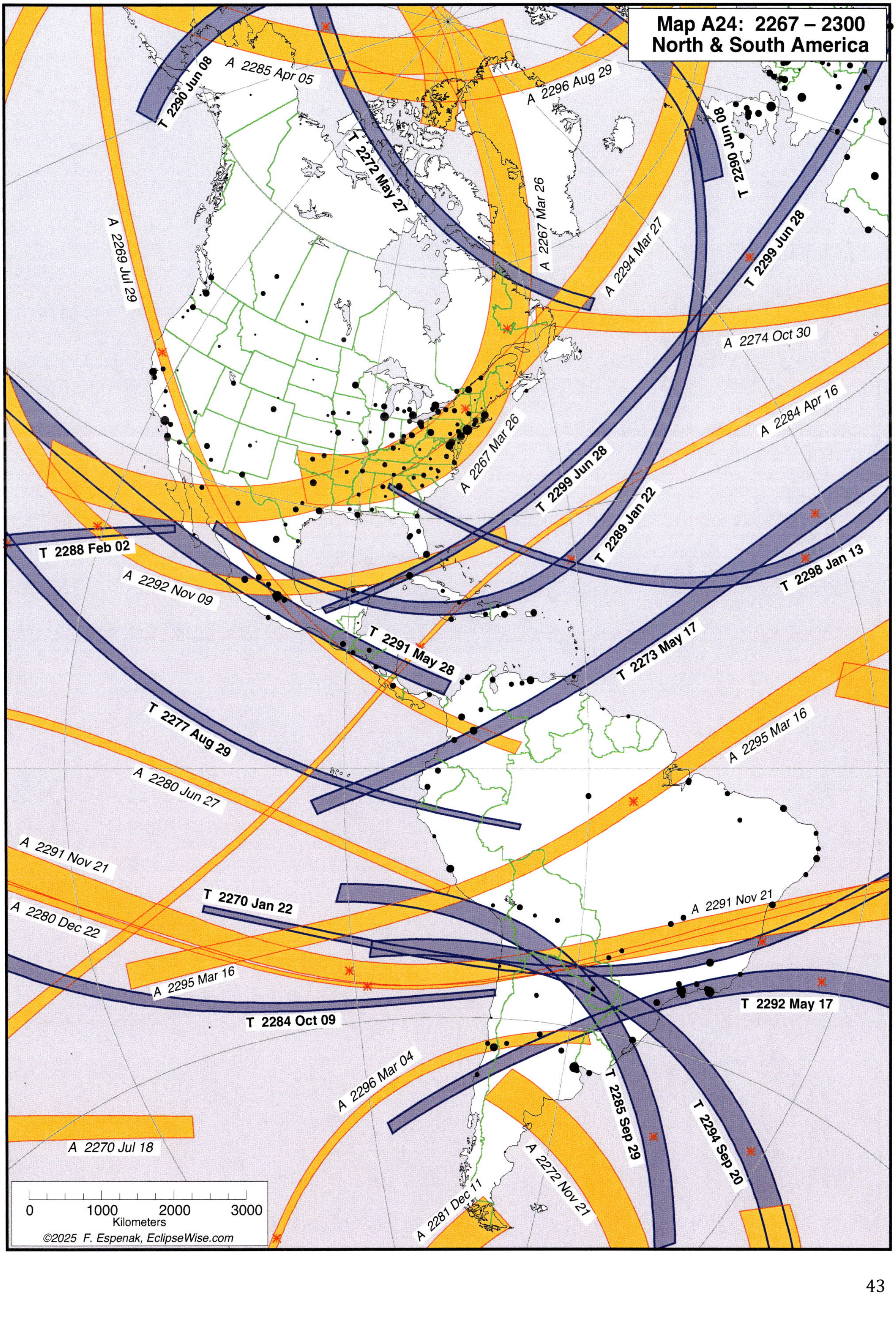
Map A24: 2267 – 2300
North & South America
T 2290 Jun 08
A 2285 Apr 05
A 2296 Aug 29
T 2272 May 27
A 2267 Mar 26
T 2290 Jun 08
A 2269 Jul 29
A 2294 Mar 27
T 2299 Jun 28
A 2274 Oct 30
A 2284 Apr 16
A 2267 Mar 26
T 2299 Jun 28
T 2289 Jan 22
T 2288 Feb 02
T 2298 Jan 13
A 2292 Nov 09
T 2291 May 28
T 2273 May 17
T 2277 Aug 29
A 2295 Mar 16
A 2280 Jun 27
A 2291 Nov 21
T 2270 Jan 22
A 2291 Nov 21
A 2280 Dec 22
A 2295 Mar 16
T 2292 May 17
T 2284 Oct 09
A 2296 Mar 04
T 2285 Sep 29
T 2294 Sep 20
A 2270 Jul 18
A 2272 Nov 21
A 2281 Dec 11
0 1000 2000 3000
Kilometers
©2025 F. Espenak, EclipseWise.com

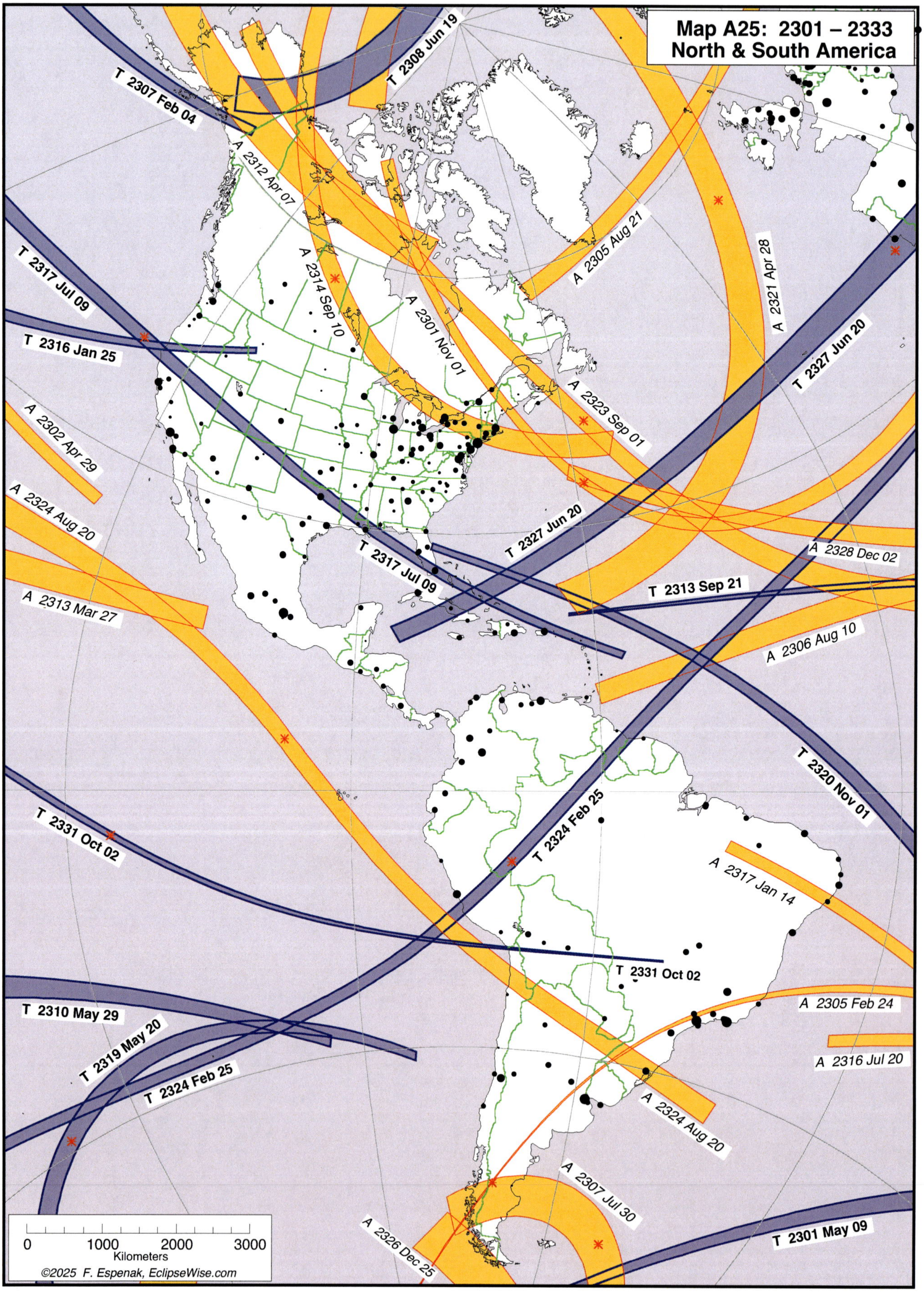

Map A25: 2301 – 2333
North & South America
T 2307 Feb 04
T 2308 Jun 19
A 2312 Apr 07
A 2305 Aug 21
A 2321 Apr 28
A 2314 Sep 10
T 2317 Jul 09
A 2301 Nov 01
T 2316 Jan 25
T 2327 Jun 20
A 2323 Sep 01
A 2302 Apr 29
A 2324 Aug 20
T 2327 Jun 20
A 2328 Dec 02
T 2317 Jul 09
T 2313 Sep 21
A 2313 Mar 27
A 2306 Aug 10
T 2320 Nov 01
T 2324 Feb 25
T 2331 Oct 02
A 2317 Jan 14
T 2331 Oct 02
A 2305 Feb 24
T 2310 May 29
T 2319 May 20
A 2316 Jul 20
T 2324 Feb 25
A 2324 Aug 20
A 2307 Jul 30
A 2326 Dec 25
T 2301 May 09
0
1000
2000
3000
Kilometers
©2025 F. Espenak, EclipseWise.com

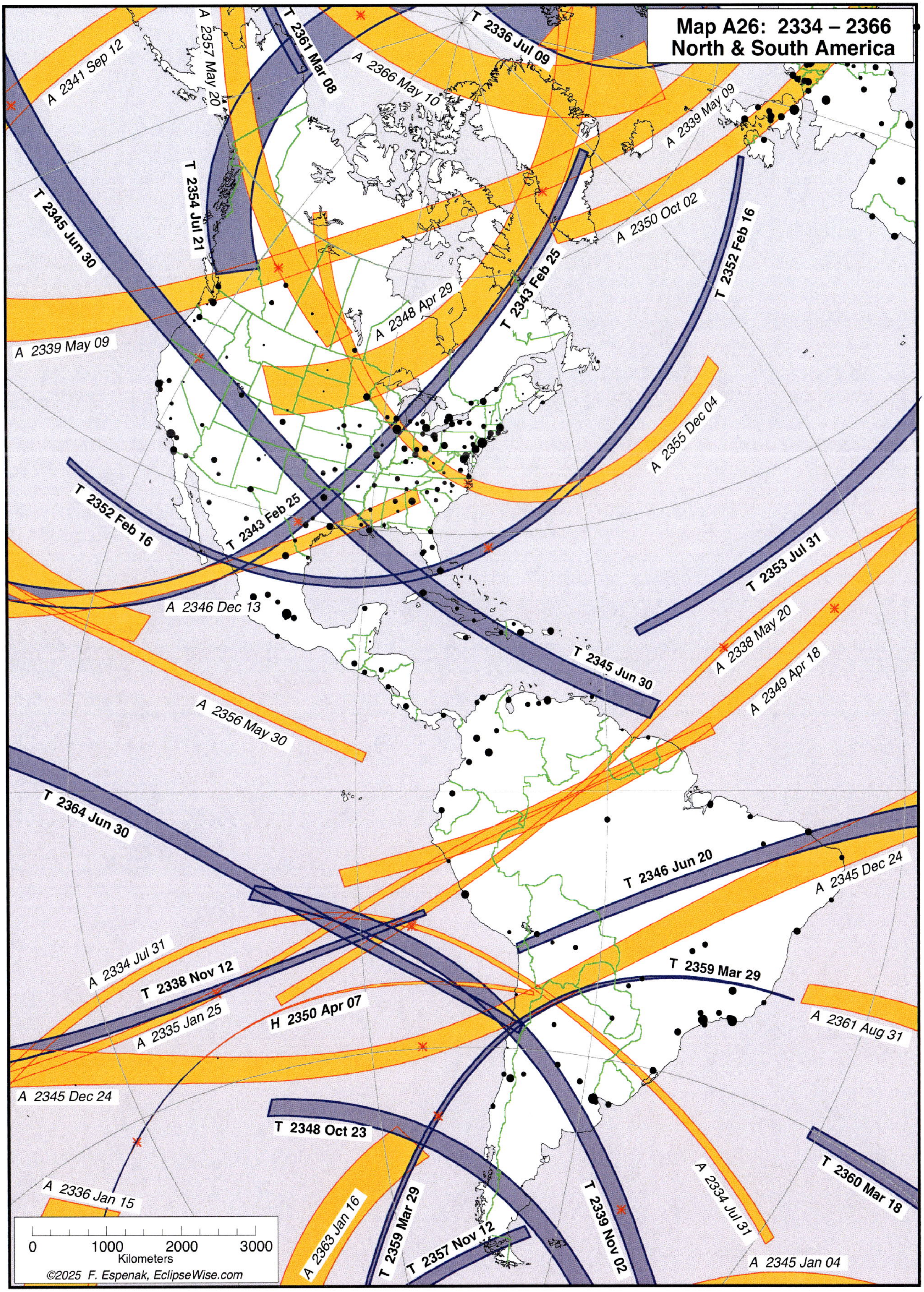
Map A26: 2334 – 2366
North & South America
T 2336 Jul 09
A 2366 May 10
T 2361 Mar 08
A 2357 May 20
A 2341 Sep 12
A 2339 May 09
T 2354 Jul 21
T 2345 Jun 30
A 2350 Oct 02
T 2352 Feb 16
T 2343 Feb 25
A 2348 Apr 29
A 2339 May 09
A 2355 Dec 04
T 2352 Feb 16
T 2343 Feb 25
A 2346 Dec 13
T 2353 Jul 31
A 2338 May 20
A 2349 Apr 18
T 2345 Jun 30
A 2356 May 30
T 2364 Jun 30
T 2346 Jun 20
A 2345 Dec 24
A 2334 Jul 31
T 2338 Nov 12
T 2359 Mar 29
A 2335 Jan 25
H 2350 Apr 07
A 2361 Aug 31
A 2345 Dec 24
T 2348 Oct 23
A 2336 Jan 15
A 2363 Jan 16
T 2359 Mar 29
T 2357 Nov 12
T 2339 Nov 02
A 2334 Jul 31
T 2360 Mar 18
A 2345 Jan 04
0 1000 2000 3000
Kilometers
©2025 F. Espenak, EclipseWise.com

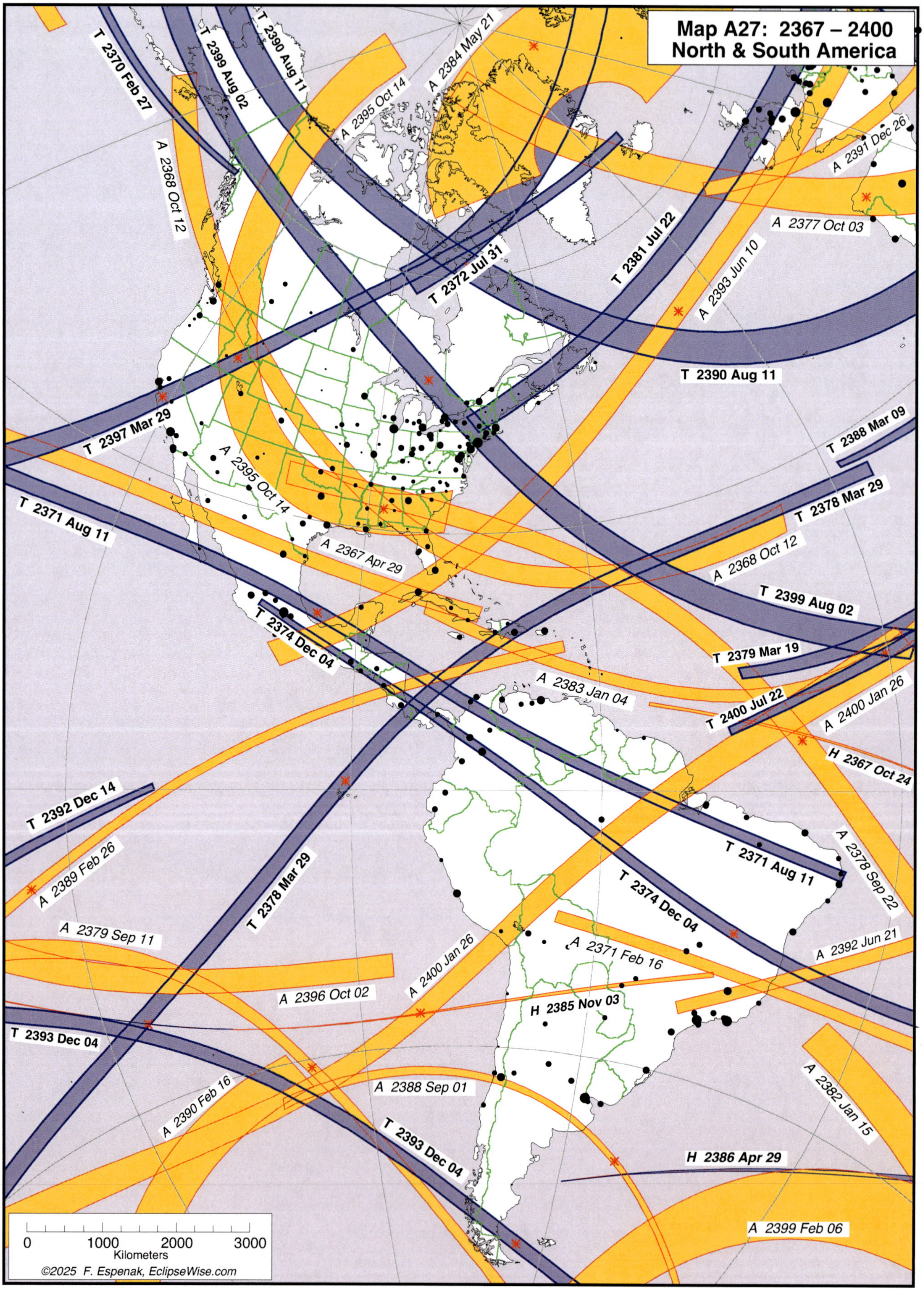
Map A27: 2367 – 2400
North & South America
T 2370 Feb 27
T 2399 Aug 02
T 2390 Aug 11
A 2384 May 21
A 2395 Oct 14
A 2368 Oct 12
A 2391 Dec 26
A 2377 Oct 03
T 2381 Jul 22
T 2372 Jul 31
A 2393 Jun 10
T 2390 Aug 11
T 2397 Mar 29
A 2395 Oct 14
T 2388 Mar 09
T 2378 Mar 29
T 2371 Aug 11
A 2367 Apr 29
A 2368 Oct 12
T 2399 Aug 02
T 2374 Dec 04
T 2379 Mar 19
A 2383 Jan 04
T 2400 Jul 22
A 2400 Jan 26
H 2367 Oct 24
T 2392 Dec 14
A 2389 Feb 26
T 2378 Mar 29
T 2371 Aug 11
A 2378 Sep 22
T 2374 Dec 04
A 2379 Sep 11
A 2371 Feb 16
A 2392 Jun 21
A 2400 Jan 26
A 2396 Oct 02
H 2385 Nov 03
T 2393 Dec 04
A 2390 Feb 16
A 2388 Sep 01
A 2382 Jan 15
T 2393 Dec 04
H 2386 Apr 29
A 2399 Feb 06
0
1000
2000
3000
Kilometers
©2025 F. Espenak, EclipseWise.com

Appendix B

World Atlas of Central Solar Eclipses

Europe & Africa: 1501 to 2400

Index for Appendix B

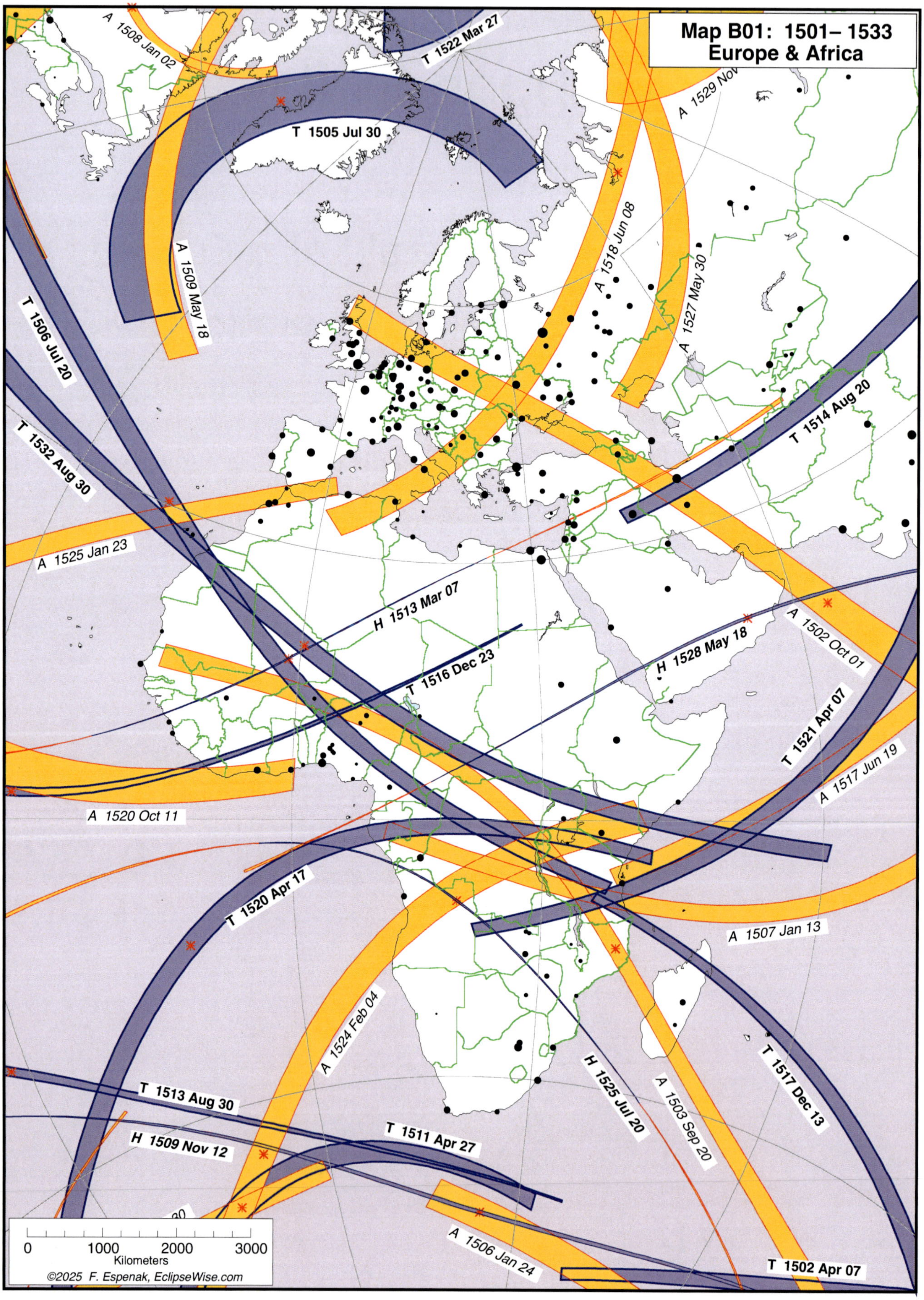
Map B01: 1501– 1533
Europe & Africa
T 1522 Mar 27
A 1508 Jan 02
T 1505 Jul 30
A 1529 Nov
A 1518 Jun 08
A 1527 May 30
A 1509 May 18
T 1506 Jul 20
T 1532 Aug 30
T 1514 Aug 20
A 1525 Jan 23
H 1513 Mar 07
A 1502 Oct 01
H 1528 May 18
T 1516 Dec 23
T 1521 Apr 07
A 1517 Jun 19
A 1520 Oct 11
T 1520 Apr 17
A 1507 Jan 13
A 1524 Feb 04
H 1525 Jul 20
A 1503 Sep 20
T 1517 Dec 13
T 1513 Aug 30
H 1509 Nov 12
T 1511 Apr 27
A 1506 Jan 24
T 1502 Apr 07
0
1000
2000
3000
Kilometers
©2025 F. Espenak, EclipseWise.com

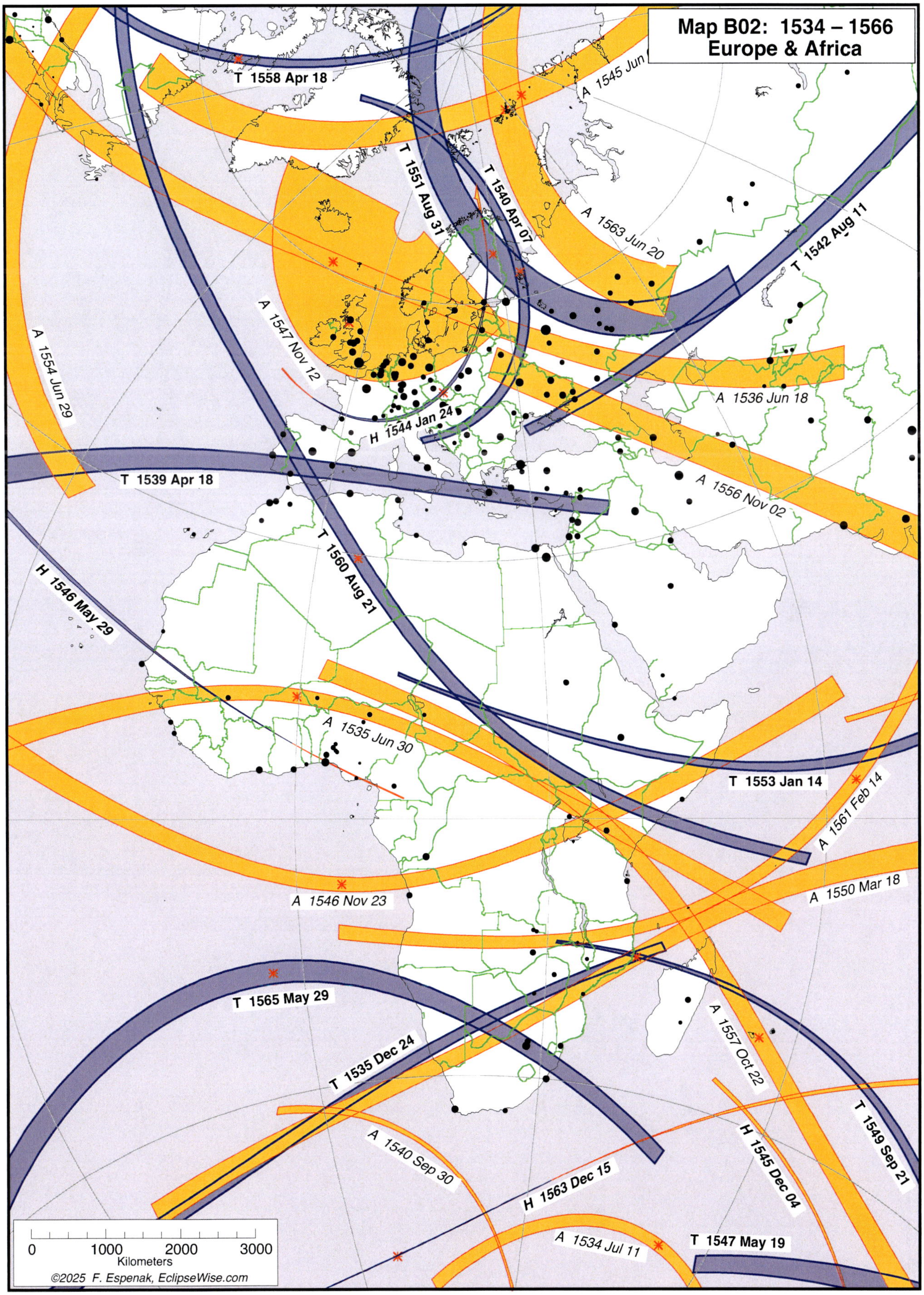
Map B02: 1534 – 1566
Europe & Africa
T 1558 Apr 18
A 1545 Jun
T 1551 Aug 31
T 1540 Apr 07
A 1563 Jun 20
T 1542 Aug 11
A 1547 Nov 12
A 1554 Jun 29
A 1536 Jun 18
H 1544 Jan 24
T 1539 Apr 18
A 1556 Nov 02
T 1560 Aug 21
H 1546 May 29
A 1535 Jun 30
T 1553 Jan 14
A 1561 Feb 14
A 1546 Nov 23
A 1550 Mar 18
T 1565 May 29
A 1557 Oct 22
T 1535 Dec 24
A 1540 Sep 30
H 1545 Dec 04
T 1549 Sep 21
H 1563 Dec 15
A 1534 Jul 11
T 1547 May 19
0 1000 2000 3000
Kilometers
©2025 F. Espenak, EclipseWise.com

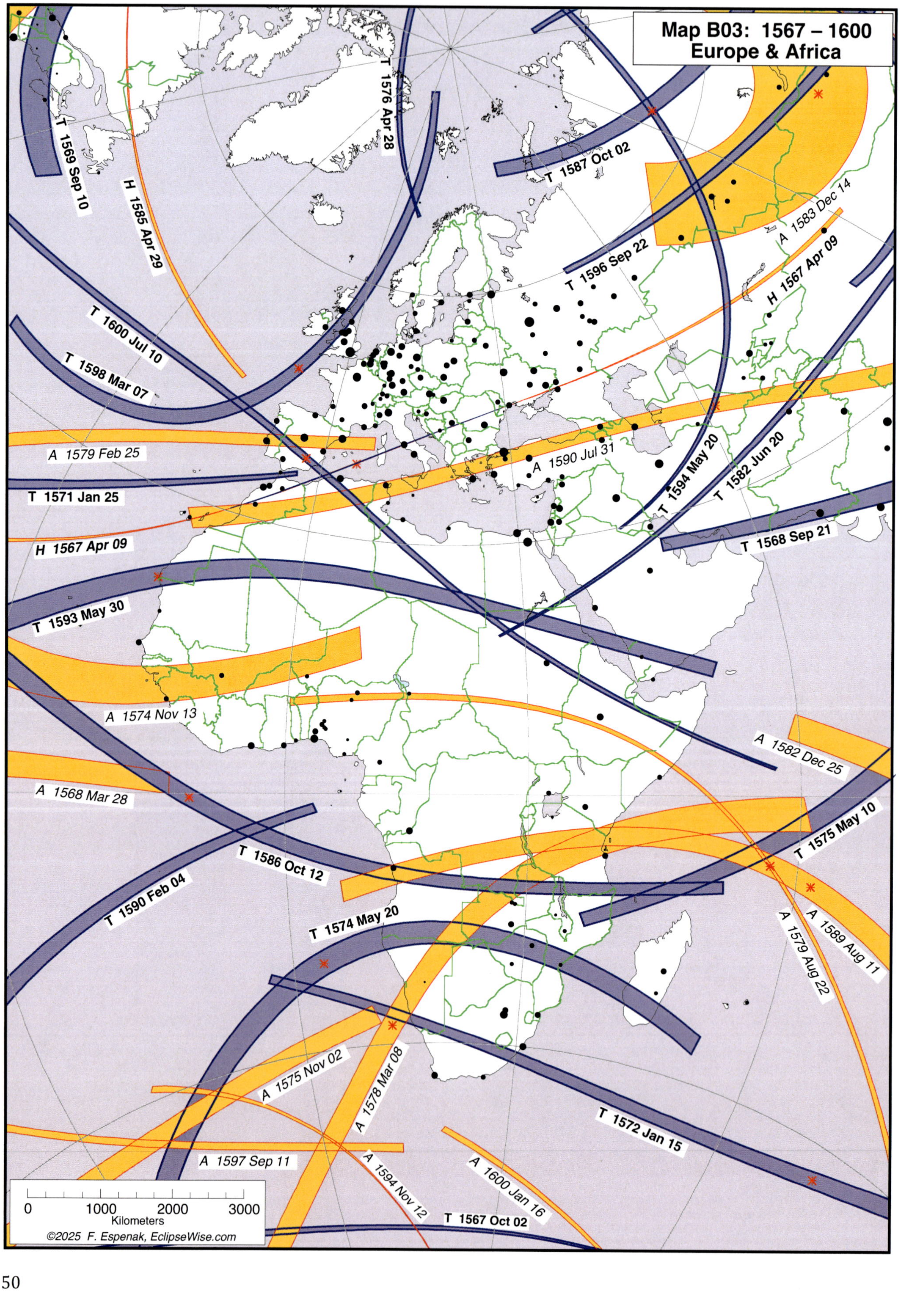
Map B03: 1567 – 1600
Europe & Africa
T 1569 Sep 10
H 1585 Apr 29
T 1576 Apr 28
T 1587 Oct 02
A 1583 Dec 14
T 1596 Sep 22
H 1567 Apr 09
T 1600 Jul 10
T 1598 Mar 07
A 1579 Feb 25
A 1590 Jul 31
T 1571 Jan 25
T 1594 May 20
T 1582 Jun 20
H 1567 Apr 09
T 1568 Sep 21
T 1593 May 30
A 1574 Nov 13
A 1582 Dec 25
A 1568 Mar 28
T 1575 May 10
T 1586 Oct 12
T 1590 Feb 04
T 1574 May 20
A 1589 Aug 11
A 1579 Aug 22
A 1575 Nov 02
A 1578 Mar 08
T 1572 Jan 15
A 1597 Sep 11
A 1594 Nov 12
A 1600 Jan 16
T 1567 Oct 02
0
1000
2000
3000
Kilometers
©2025 F. Espenak, EclipseWise.com

Map B04: 1601 – 1633
Europe & Africa
T 1616 Mar 17
H 1630 Jun 10
T 1605 Oct 12
H 1627 Aug 11
A 1601 Dec 24
T 1625 Mar 08
A 1621 May 21
T 1612 May 30
A 1601 Dec 24
T 1607 Feb 26
A 1608 Aug 10
A 1633 Apr 08
A 1628 Dec 25
A 1610 Dec 15
A 1622 May 10
A 1633 Oct 03
T 1622 Nov 03
T 1614 Oct 03
T 1619 Jul 11
A 1601 Jan 04
A 1615 Mar 29
T 1629 Jun 21
T 1626 Feb 26
A 1629 Dec 14
A 1607 Aug 22
T 1626 Feb 26
A 1623 Apr 29
A 1632 Apr 19
A 1611 Dec 04
T 1628 Jul 01
T 1603 Nov 03
A 1618 Jan 26
T 1621 Nov 13
0 1000 2000 3000
Kilometers
©2025 F. Espenak, EclipseWise.com

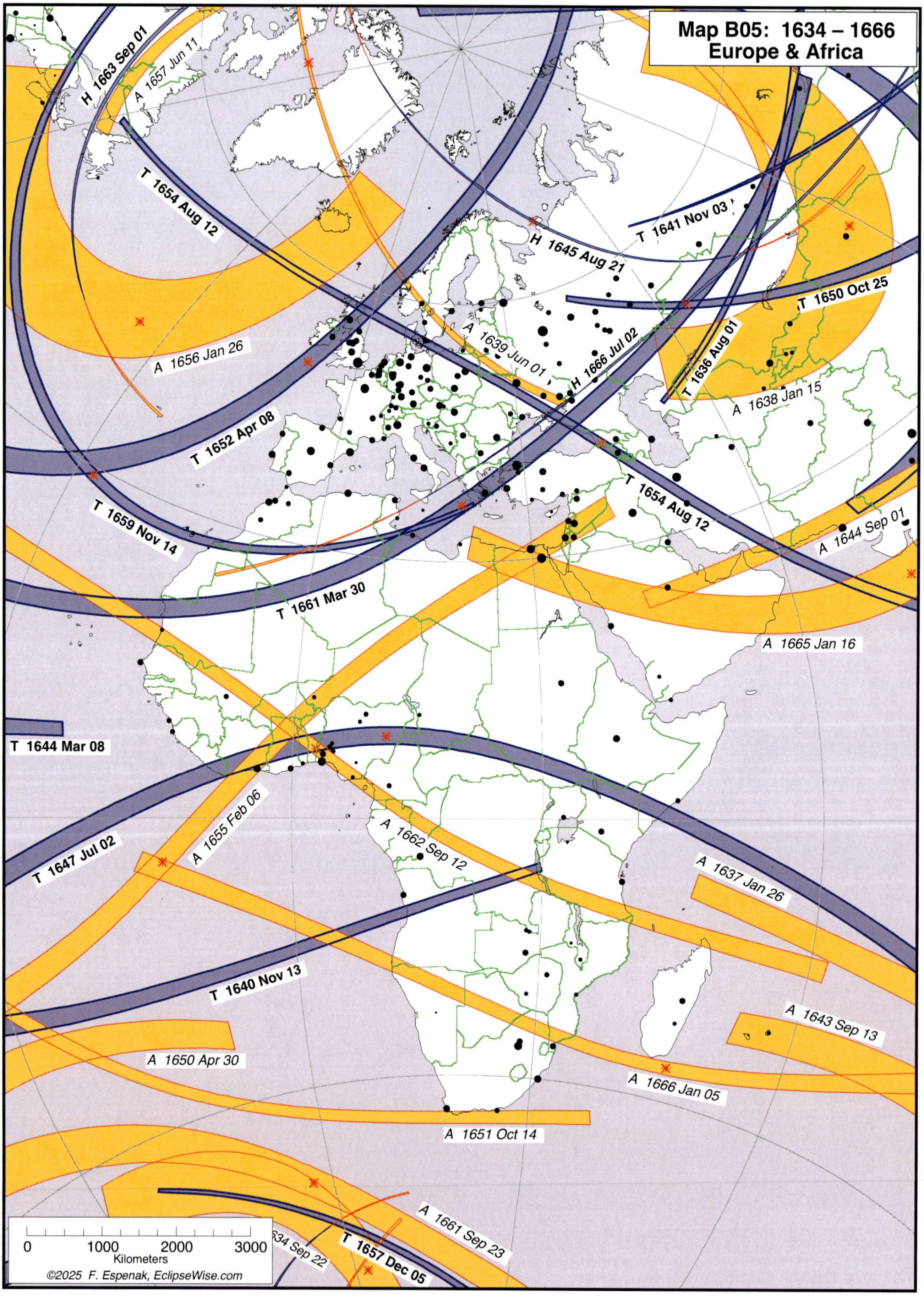
Map B05: 1634 – 1666
Europe & Africa
H 1663 Sep 01
A 1657 Jun 11
T 1654 Aug 12
A 1656 Jan 26
T 1652 Apr 08
T 1659 Nov 14
H 1645 Aug 21
T 1641 Nov 03
T 1650 Oct 25
A 1639 Jun 01
H 1666 Jul 02
T 1636 Aug 01
A 1638 Jan 15
T 1654 Aug 12
A 1644 Sep 01
T 1661 Mar 30
A 1665 Jan 16
T 1644 Mar 08
A 1655 Feb 06
T 1647 Jul 02
A 1662 Sep 12
A 1637 Jan 26
T 1640 Nov 13
A 1643 Sep 13
A 1650 Apr 30
A 1666 Jan 05
A 1651 Oct 14
A 1661 Sep 23
T 1657 Dec 05
0 1000 2000 3000
Kilometers
©2025 F. Espenak, EclipseWise.com

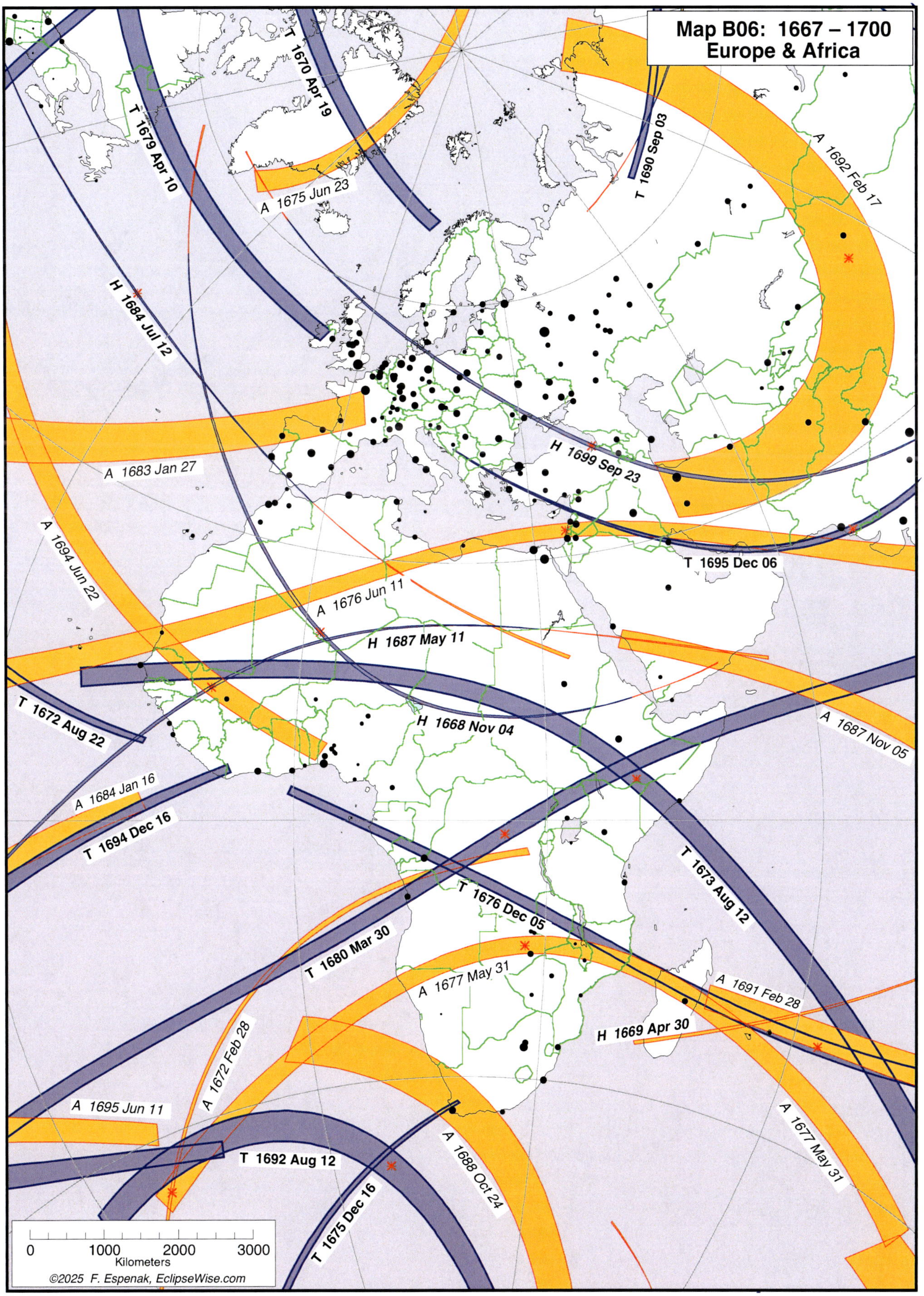
Map B06: 1667 – 1700
Europe & Africa
T 1670 Apr 19
T 1679 Apr 10
A 1675 Jun 23
T 1690 Sep 03
A 1692 Feb 17
H 1684 Jul 12
A 1683 Jan 27
H 1699 Sep 23
A 1694 Jun 22
T 1695 Dec 06
A 1676 Jun 11
H 1687 May 11
T 1672 Aug 22
H 1668 Nov 04
A 1687 Nov 05
A 1684 Jan 16
T 1694 Dec 16
T 1673 Aug 12
T 1676 Dec 05
T 1680 Mar 30
A 1677 May 31
A 1691 Feb 28
H 1669 Apr 30
A 1672 Feb 28
A 1695 Jun 11
T 1692 Aug 12
A 1688 Oct 24
A 1677 May 31
T 1675 Dec 16
0
1000
2000
3000
Kilometers
©2025 F. Espenak, EclipseWise.com

Map B07: 1701 – 1733
Europe & Africa
A 1728 Mar 10
T 1733 May 13
T 1724 May 22
T 1726 Sep 25
A 1710 Feb 28
T 1715 May 03
T 1706 May 12
A 1730 Jul 15
A 1704 Nov 27
T 1708 Sep 14
A 1709 Mar 11
A 1720 Aug 04
H 1717 Oct 04
T 1727 Sep 15
A 1719 Feb 19
T 1726 Sep 25
A 1722 Dec 08
A 1716 Oct 15
A 1731 Jul 04
A 1709 Mar 11
T 1701 Aug 04
A 1726 Apr 02
A 1720 Feb 08
T 1727 Sep 15
T 1731 Jan 08
A 1716 Oct 15
T 1730 Jan 18
A 1705 Nov 16
H 1717 Apr 11
A 1726 Apr 02
T 1701 Aug 04
A 1704 Jun 02
27
0 1000 2000 3000
Kilometers
©2025 F. Espenak, EclipseWise.com

Map B08: 1734 – 1766
Europe & Africa
A 1739 Aug 04
A 1757 Aug 14
H 1735 Oct 16
T 1751 May 25
T 1760 Jun 13
A 1766 Aug 05
A 1748 Jul 25
A 1737 Mar 01
T 1762 Oct 17
A 1738 Aug 15
A 1764 Apr 01
A 1763 Apr 13
H 1753 Oct 26
A 1758 Dec 30
H 1750 Jan 08
A 1766 Aug 05
T 1752 May 13
T 1734 May 03
A 1748 Jul 25
T 1760 Jun 13
A 1763 Apr 13
A 1738 Aug 15
T 1741 Jun 13
A 1749 Jul 14
A 1741 Dec 08
T 1734 May 03
T 1741 Jun 13
A 1759 Dec 19
T 1759 Jun 24
T 1746 Sep 15
T 1755 Sep 06
T 1766 Feb 09
A 1742 Nov 27
T 1745 Sep 25
H 1753 May 03
T 1757 Feb 18
T 1766 Feb 09
0 1000 2000 3000
Kilometers
©2025 F. Espenak, EclipseWise.com

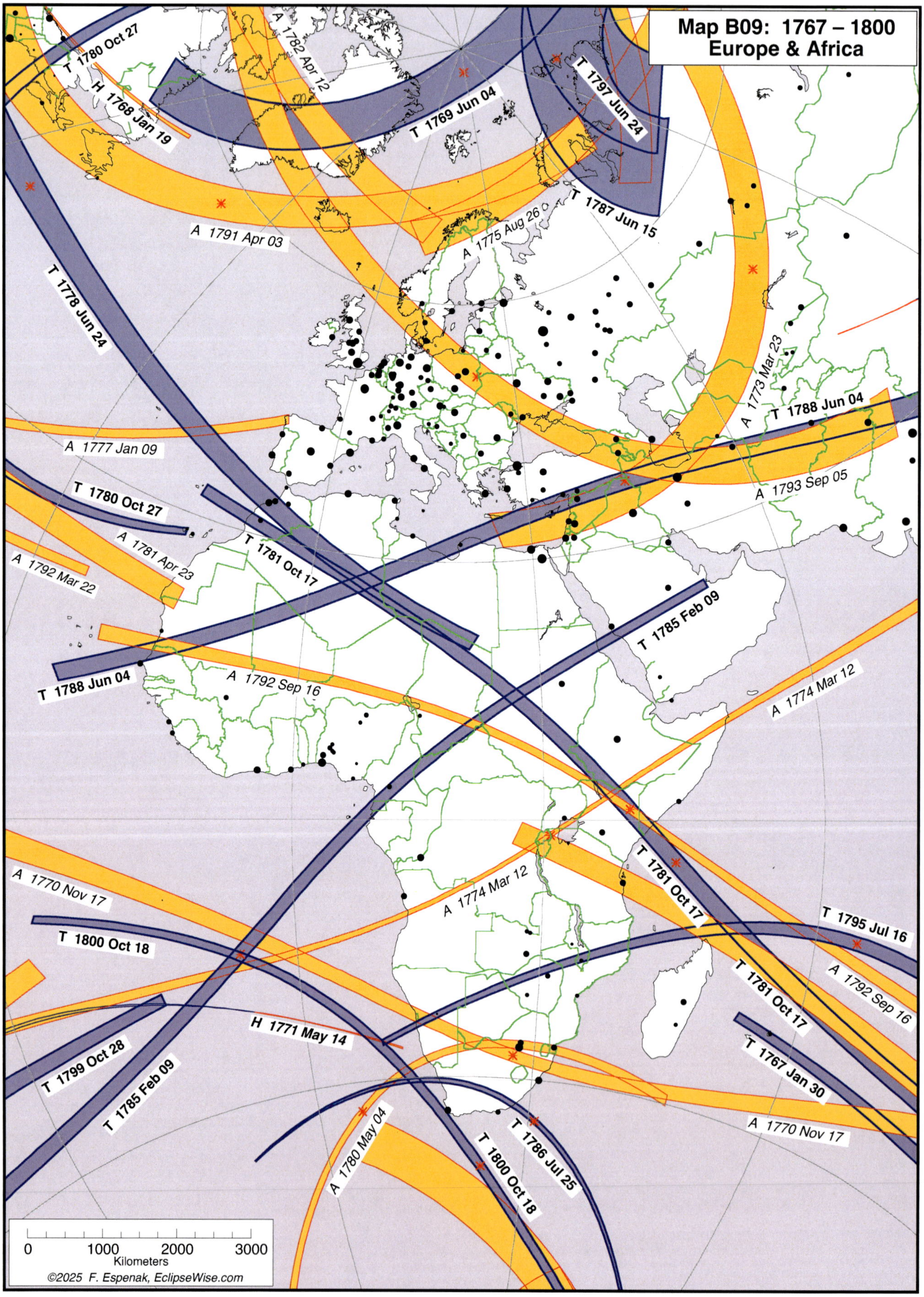
Map B09: 1767 – 1800
Europe & Africa
T 1780 Oct 27
H 1768 Jan 19
A 1782 Apr 12
T 1769 Jun 04
T 1797 Jun 24
T 1787 Jun 15
A 1791 Apr 03
A 1775 Aug 26
T 1778 Jun 24
A 1773 Mar 23
T 1788 Jun 04
A 1777 Jan 09
A 1793 Sep 05
T 1780 Oct 27
A 1781 Apr 23
A 1792 Mar 22
T 1781 Oct 17
T 1785 Feb 09
T 1788 Jun 04
A 1792 Sep 16
A 1774 Mar 12
A 1770 Nov 17
A 1774 Mar 12
T 1781 Oct 17
T 1795 Jul 16
T 1800 Oct 18
A 1792 Sep 16
T 1781 Oct 17
H 1771 May 14
T 1767 Jan 30
T 1799 Oct 28
T 1785 Feb 09
A 1770 Nov 17
A 1780 May 04
T 1786 Jul 25
T 1800 Oct 18
0
1000
2000
3000
Kilometers
©2025 F. Espenak, EclipseWise.com

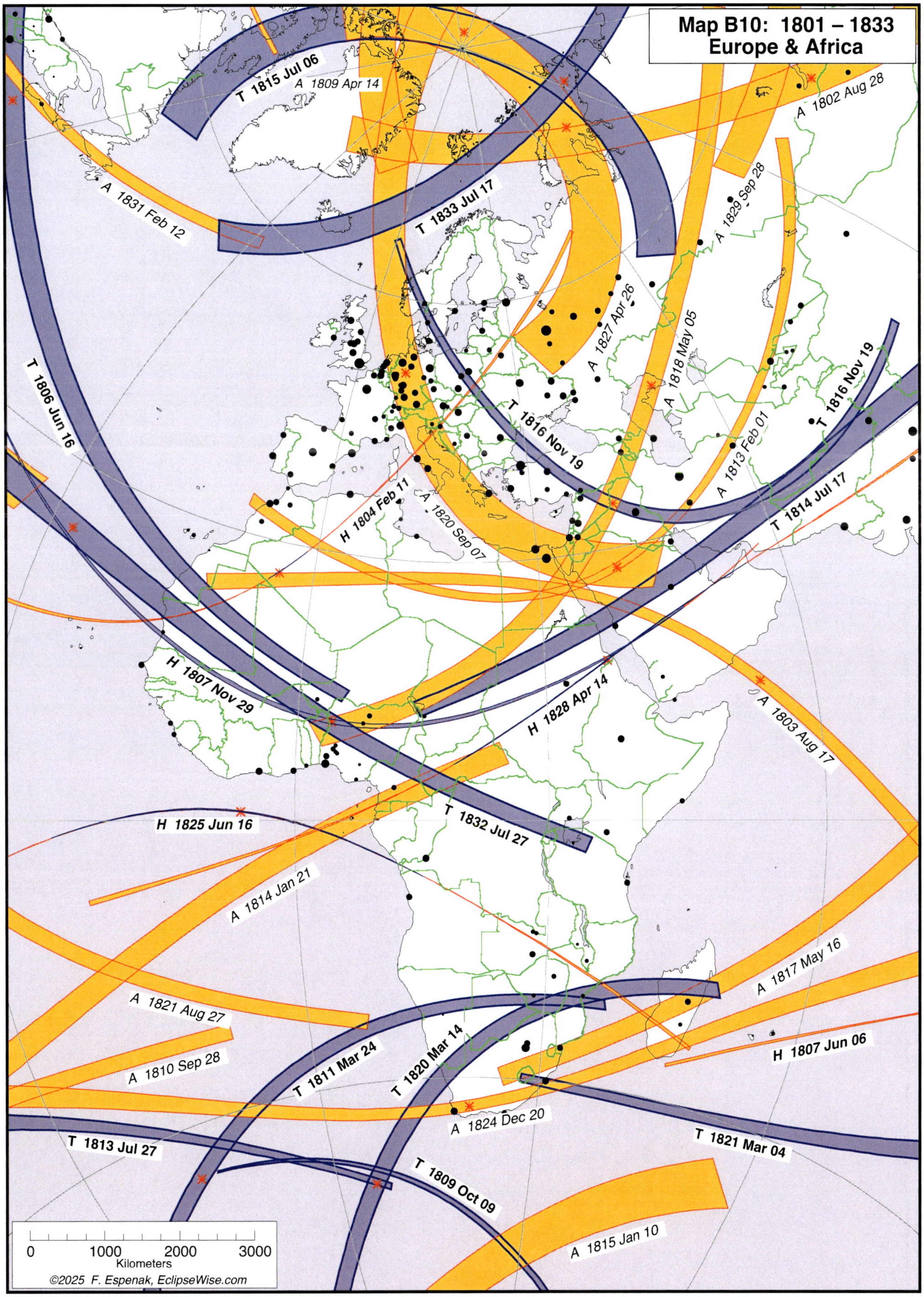
Map B10: 1801 – 1833
Europe & Africa
T 1815 Jul 06
A 1809 Apr 14
A 1802 Aug 28
A 1831 Feb 12
T 1833 Jul 17
A 1829 Sep 28
A 1827 Apr 26
A 1818 May 05
T 1806 Jun 16
T 1816 Nov 19
T 1816 Nov 19
A 1813 Feb 01
H 1804 Feb 11
A 1820 Sep 07
T 1814 Jul 17
H 1807 Nov 29
H 1828 Apr 14
A 1803 Aug 17
T 1832 Jul 27
H 1825 Jun 16
A 1814 Jan 21
A 1817 May 16
A 1821 Aug 27
A 1810 Sep 28
T 1811 Mar 24
T 1820 Mar 14
H 1807 Jun 06
A 1824 Dec 20
T 1813 Jul 27
T 1821 Mar 04
T 1809 Oct 09
A 1815 Jan 10
0 1000 2000 3000
Kilometers
©2025 F. Espenak, EclipseWise.com

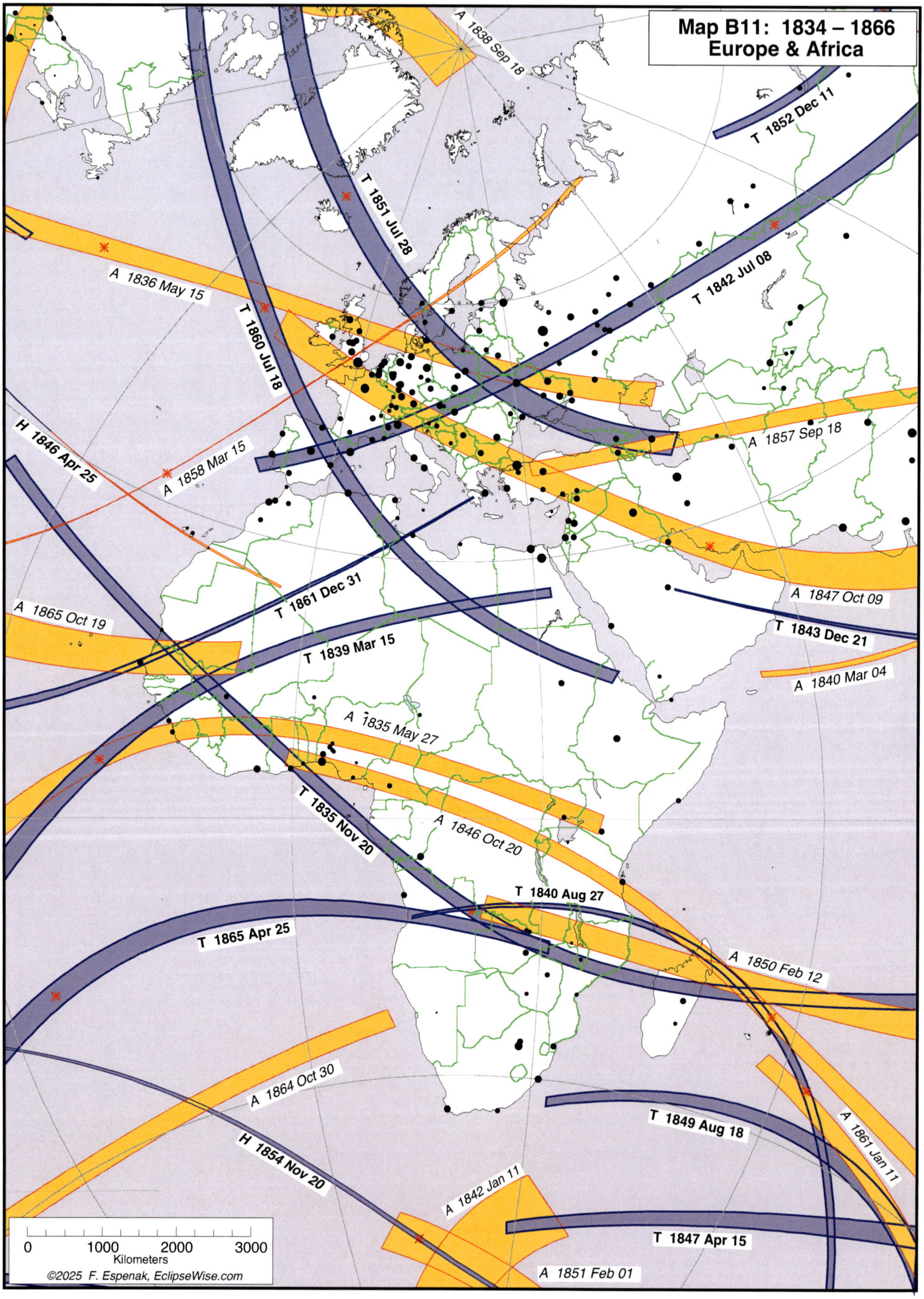
Map B11: 1834 – 1866
Europe & Africa
A 1838 Sep 18
T 1852 Dec 11
T 1851 Jul 28
T 1842 Jul 08
A 1836 May 15
T 1860 Jul 18
A 1857 Sep 18
H 1846 Apr 25
A 1858 Mar 15
T 1861 Dec 31
A 1847 Oct 09
A 1865 Oct 19
T 1843 Dec 21
T 1839 Mar 15
A 1840 Mar 04
A 1835 May 27
T 1835 Nov 20
A 1846 Oct 20
T 1840 Aug 27
T 1865 Apr 25
A 1850 Feb 12
A 1864 Oct 30
T 1849 Aug 18
A 1861 Jan 11
H 1854 Nov 20
A 1842 Jan 11
T 1847 Apr 15
A 1851 Feb 01
0 1000 2000 3000
Kilometers
©2025 F. Espenak, EclipseWise.com

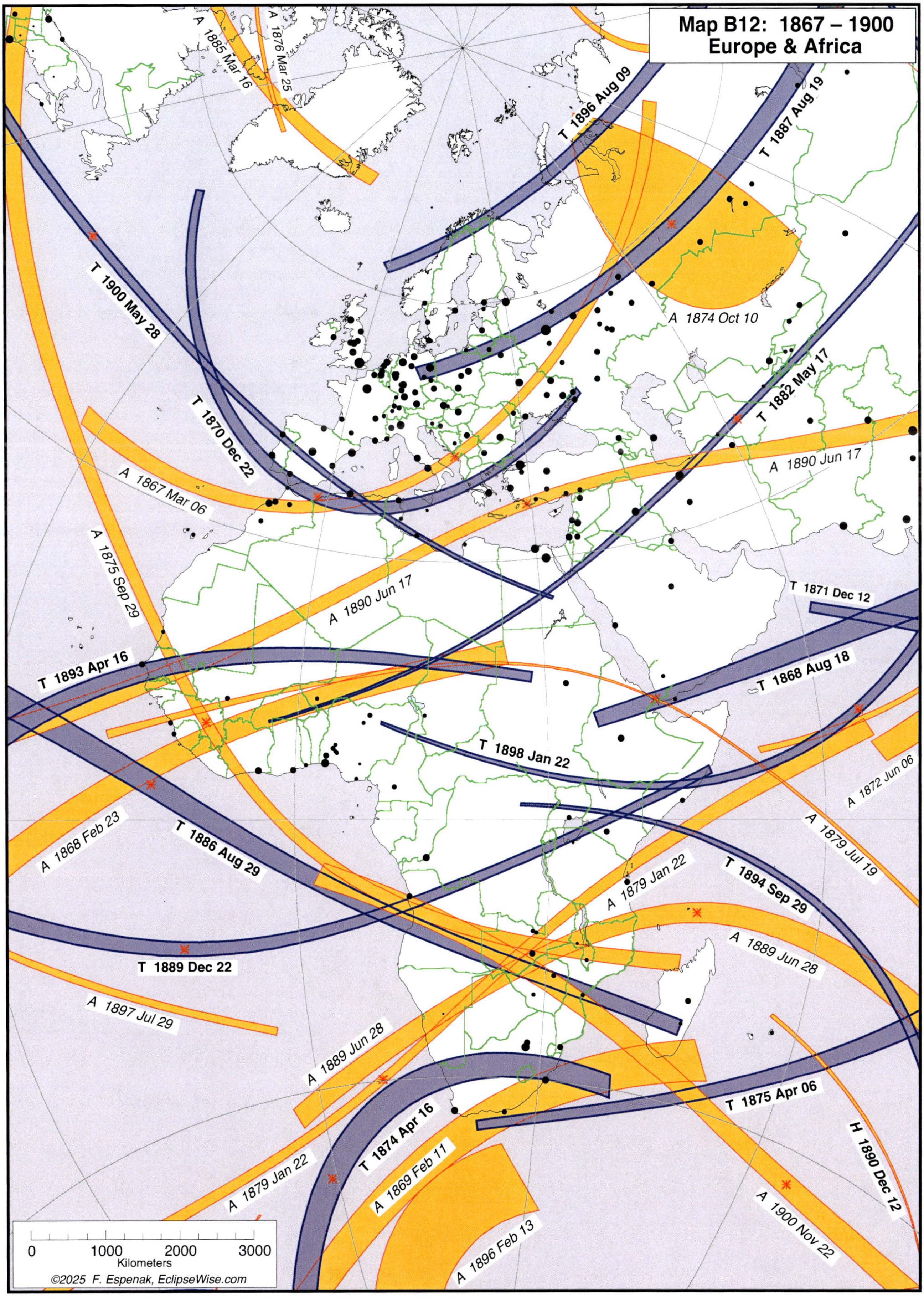
Map B12: 1867 – 1900
Europe & Africa
A 1885 Mar 16
A 1876 Mar 25
T 1896 Aug 09
T 1887 Aug 19
A 1874 Oct 10
T 1900 May 28
T 1870 Dec 22
A 1867 Mar 06
T 1882 May 17
A 1890 Jun 17
A 1875 Sep 29
A 1890 Jun 17
T 1871 Dec 12
T 1893 Apr 16
T 1868 Aug 18
T 1898 Jan 22
A 1872 Jun 06
A 1868 Feb 23
T 1886 Aug 29
A 1879 Jul 19
A 1879 Jan 22
T 1894 Sep 29
A 1889 Jun 28
T 1889 Dec 22
A 1897 Jul 29
A 1889 Jun 28
T 1875 Apr 06
T 1874 Apr 16
H 1890 Dec 12
A 1869 Feb 11
A 1879 Jan 22
A 1900 Nov 22
A 1896 Feb 13
0
1000
2000
3000
Kilometers
©2025 F. Espenak, EclipseWise.com

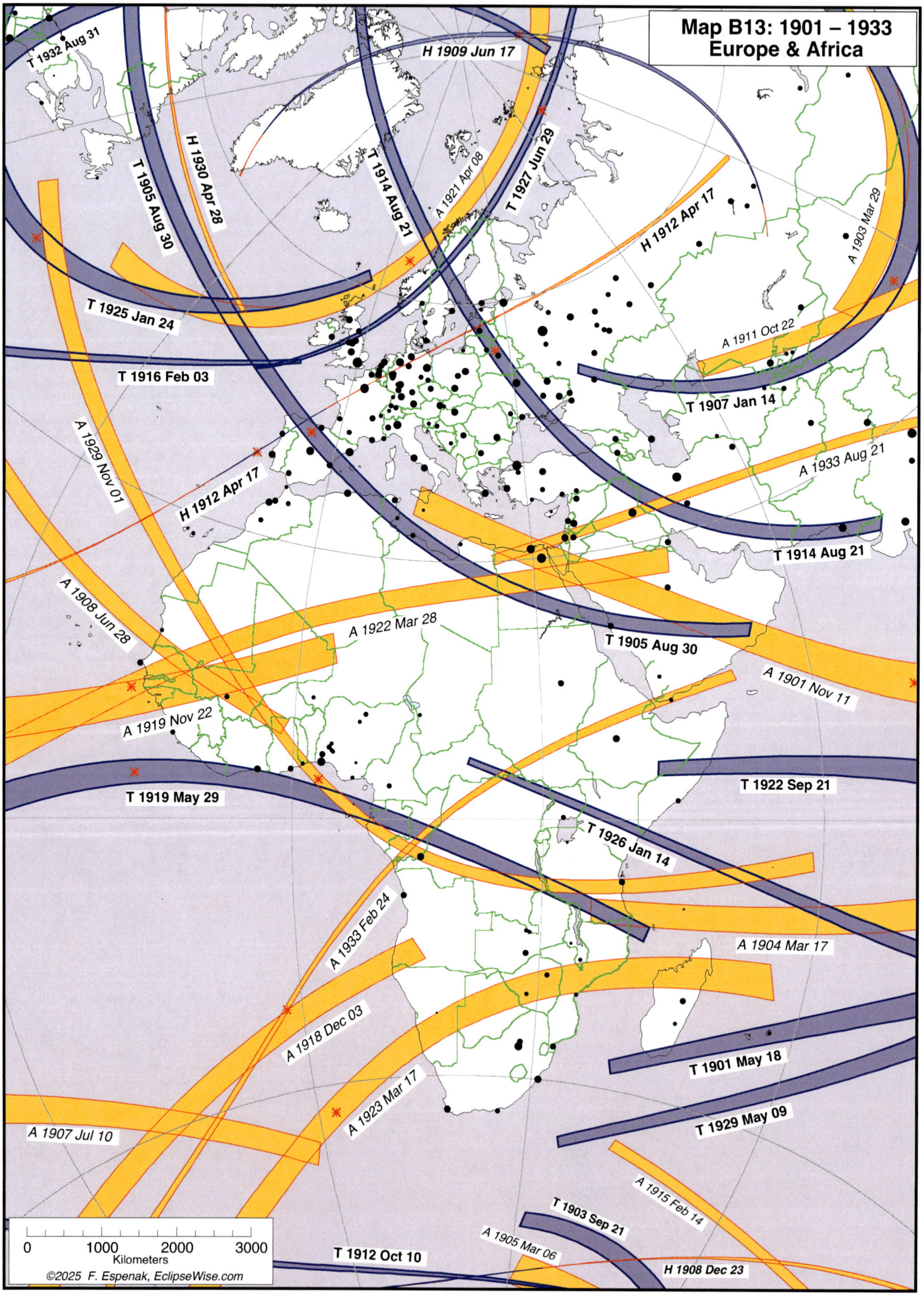
Map B13: 1901 – 1933
Europe & Africa
T 1932 Aug 31
H 1909 Jun 17
T 1905 Aug 30
H 1930 Apr 28
T 1914 Aug 21
A 1921 Apr 08
T 1927 Jun 29
H 1912 Apr 17
A 1903 Mar 29
T 1925 Jan 24
A 1911 Oct 22
T 1916 Feb 03
T 1907 Jan 14
A 1929 Nov 01
A 1933 Aug 21
H 1912 Apr 17
T 1914 Aug 21
A 1908 Jun 28
A 1922 Mar 28
T 1905 Aug 30
A 1901 Nov 11
A 1919 Nov 22
T 1922 Sep 21
T 1919 May 29
T 1926 Jan 14
A 1933 Feb 24
A 1904 Mar 17
A 1918 Dec 03
T 1901 May 18
A 1923 Mar 17
T 1929 May 09
A 1907 Jul 10
A 1915 Feb 14
T 1903 Sep 21
A 1905 Mar 06
T 1912 Oct 10
H 1908 Dec 23
0
1000
2000
3000
Kilometers
©2025 F. Espenak, EclipseWise.com

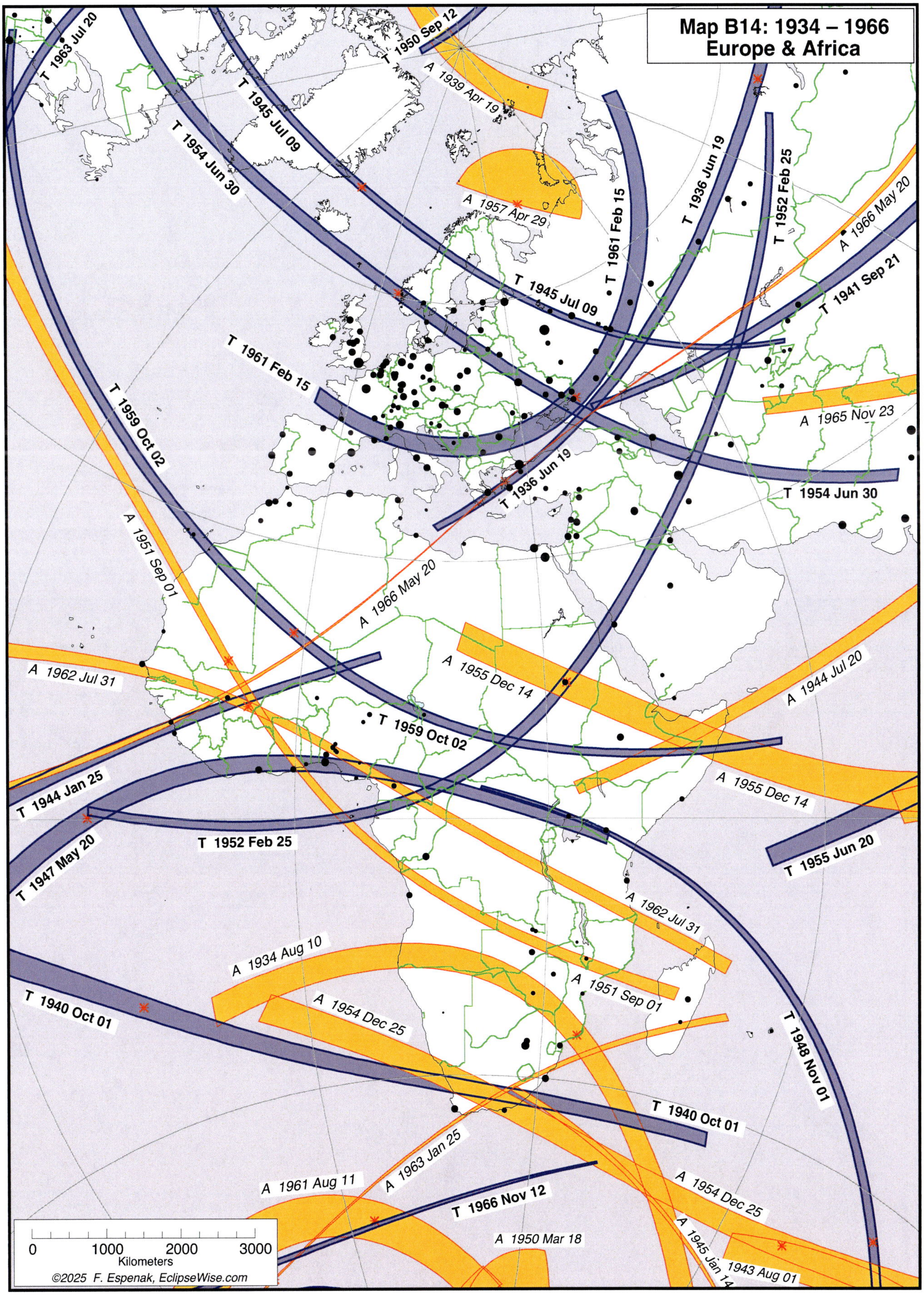
Map B14: 1934 – 1966
Europe & Africa
T 1963 Jul 20
T 1950 Sep 12
A 1939 Apr 19
T 1945 Jul 09
T 1954 Jun 30
A 1957 Apr 29
T 1961 Feb 15
T 1936 Jun 19
T 1952 Feb 25
A 1966 May 20
T 1941 Sep 21
A 1965 Nov 23
T 1959 Oct 02
A 1951 Sep 01
A 1962 Jul 31
A 1955 Dec 14
A 1944 Jul 20
T 1944 Jan 25
T 1947 May 20
T 1955 Jun 20
T 1948 Nov 01
A 1934 Aug 10
A 1954 Dec 25
T 1940 Oct 01
A 1963 Jan 25
A 1961 Aug 11
T 1966 Nov 12
A 1950 Mar 18
A 1945 Jan 14
A 1943 Aug 01
0 1000 2000 3000
Kilometers
©2025 F. Espenak, EclipseWise.com

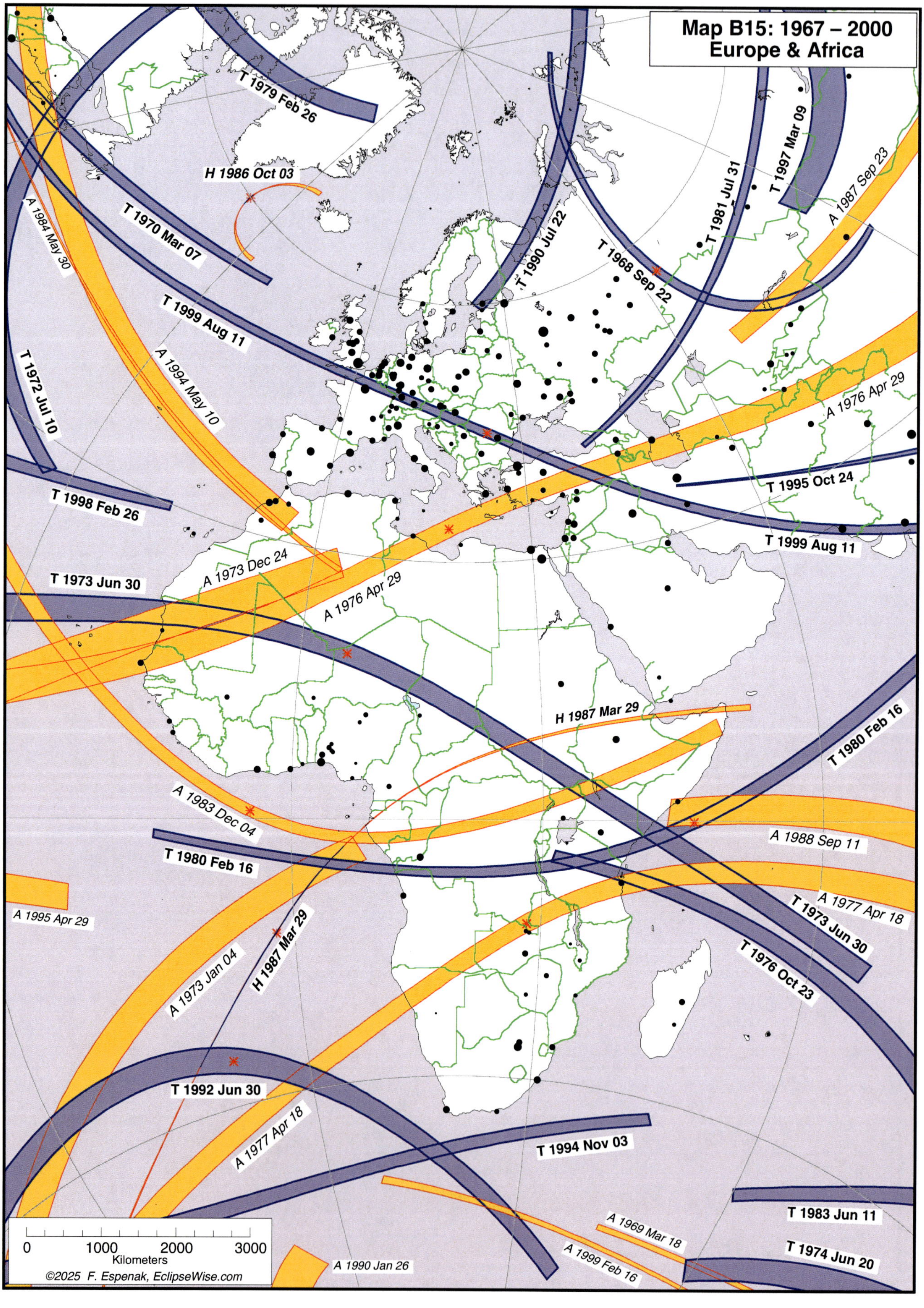

Map B15: 1967 – 2000
Europe & Africa
T 1979 Feb 26
H 1986 Oct 03
T 1970 Mar 07
A 1984 May 30
T 1999 Aug 11
A 1994 May 10
T 1972 Jul 10
T 1998 Feb 26
T 1990 Jul 22
T 1968 Sep 22
T 1981 Jul 31
T 1997 Mar 09
A 1987 Sep 23
A 1976 Apr 29
T 1995 Oct 24
T 1999 Aug 11
A 1973 Dec 24
T 1973 Jun 30
A 1976 Apr 29
H 1987 Mar 29
T 1980 Feb 16
A 1983 Dec 04
T 1980 Feb 16
A 1988 Sep 11
A 1977 Apr 18
T 1973 Jun 30
T 1976 Oct 23
A 1995 Apr 29
A 1973 Jan 04
H 1987 Mar 29
T 1992 Jun 30
A 1977 Apr 18
T 1994 Nov 03
T 1983 Jun 11
A 1969 Mar 18
T 1974 Jun 20
A 1999 Feb 16
A 1990 Jan 26
0
1000
2000
3000
Kilometers
©2025 F. Espenak, EclipseWise.com

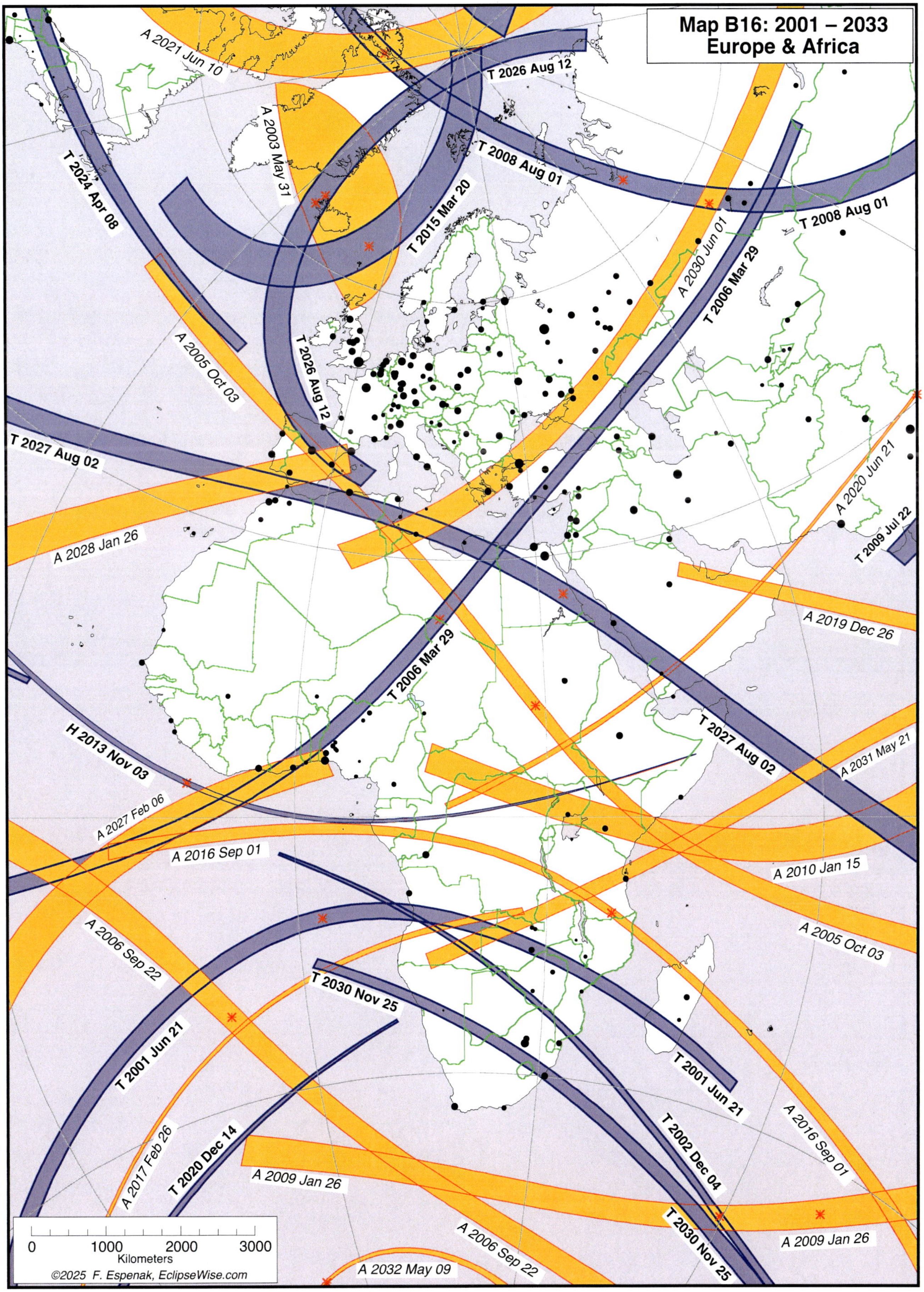
Map B16: 2001 – 2033
Europe & Africa
A 2021 Jun 10
T 2026 Aug 12
A 2003 May 31
T 2008 Aug 01
T 2024 Apr 08
T 2015 Mar 20
T 2008 Aug 01
A 2030 Jun 01
T 2006 Mar 29
A 2005 Oct 03
T 2026 Aug 12
T 2027 Aug 02
A 2028 Jan 26
A 2020 Jun 21
T 2009 Jul 22
A 2019 Dec 26
T 2006 Mar 29
H 2013 Nov 03
T 2027 Aug 02
A 2031 May 21
A 2027 Feb 06
A 2016 Sep 01
A 2010 Jan 15
A 2006 Sep 22
A 2005 Oct 03
T 2030 Nov 25
T 2001 Jun 21
T 2001 Jun 21
A 2016 Sep 01
A 2017 Feb 26
T 2020 Dec 14
A 2009 Jan 26
T 2002 Dec 04
T 2030 Nov 25
A 2009 Jan 26
A 2006 Sep 22
A 2032 May 09
0
1000
2000
3000
Kilometers
©2025 F. Espenak, EclipseWise.com

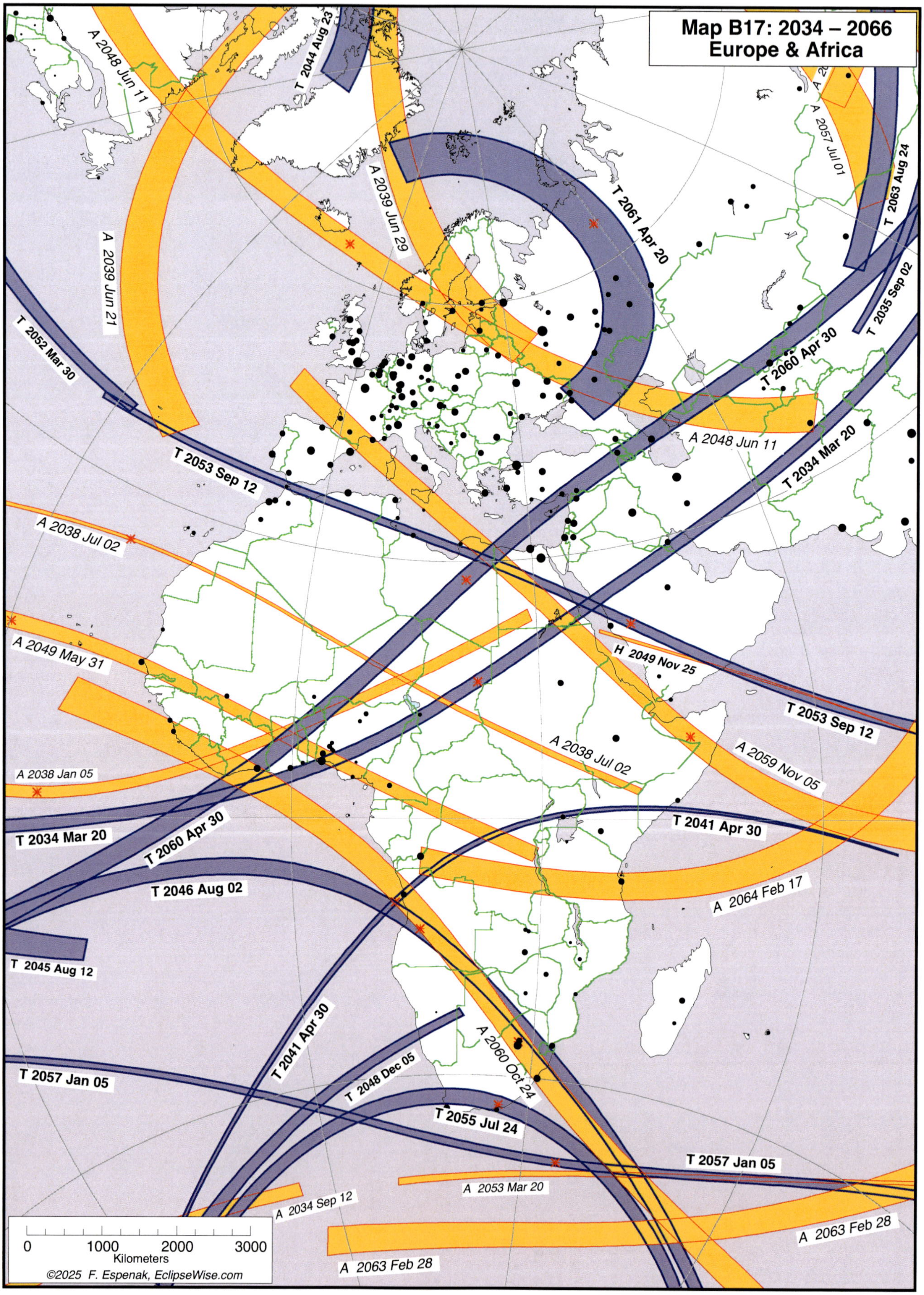

Map B17: 2034 – 2066
Europe & Africa
A 2048 Jun 11
T 2044 Aug 23
A 2057 Jul 01
T 2063 Aug 24
A 2039 Jun 29
T 2061 Apr 20
A 2039 Jun 21
T 2052 Mar 30
T 2035 Sep 02
T 2060 Apr 30
A 2048 Jun 11
T 2034 Mar 20
T 2053 Sep 12
A 2038 Jul 02
A 2049 May 31
H 2049 Nov 25
T 2053 Sep 12
A 2059 Nov 05
A 2038 Jul 02
A 2038 Jan 05
T 2041 Apr 30
T 2034 Mar 20
T 2060 Apr 30
T 2046 Aug 02
A 2064 Feb 17
T 2045 Aug 12
T 2041 Apr 30
A 2060 Oct 24
T 2048 Dec 05
T 2057 Jan 05
T 2055 Jul 24
T 2057 Jan 05
A 2053 Mar 20
A 2034 Sep 12
A 2063 Feb 28
A 2063 Feb 28
0 1000 2000 3000
Kilometers
©2025 F. Espenak, EclipseWise.com

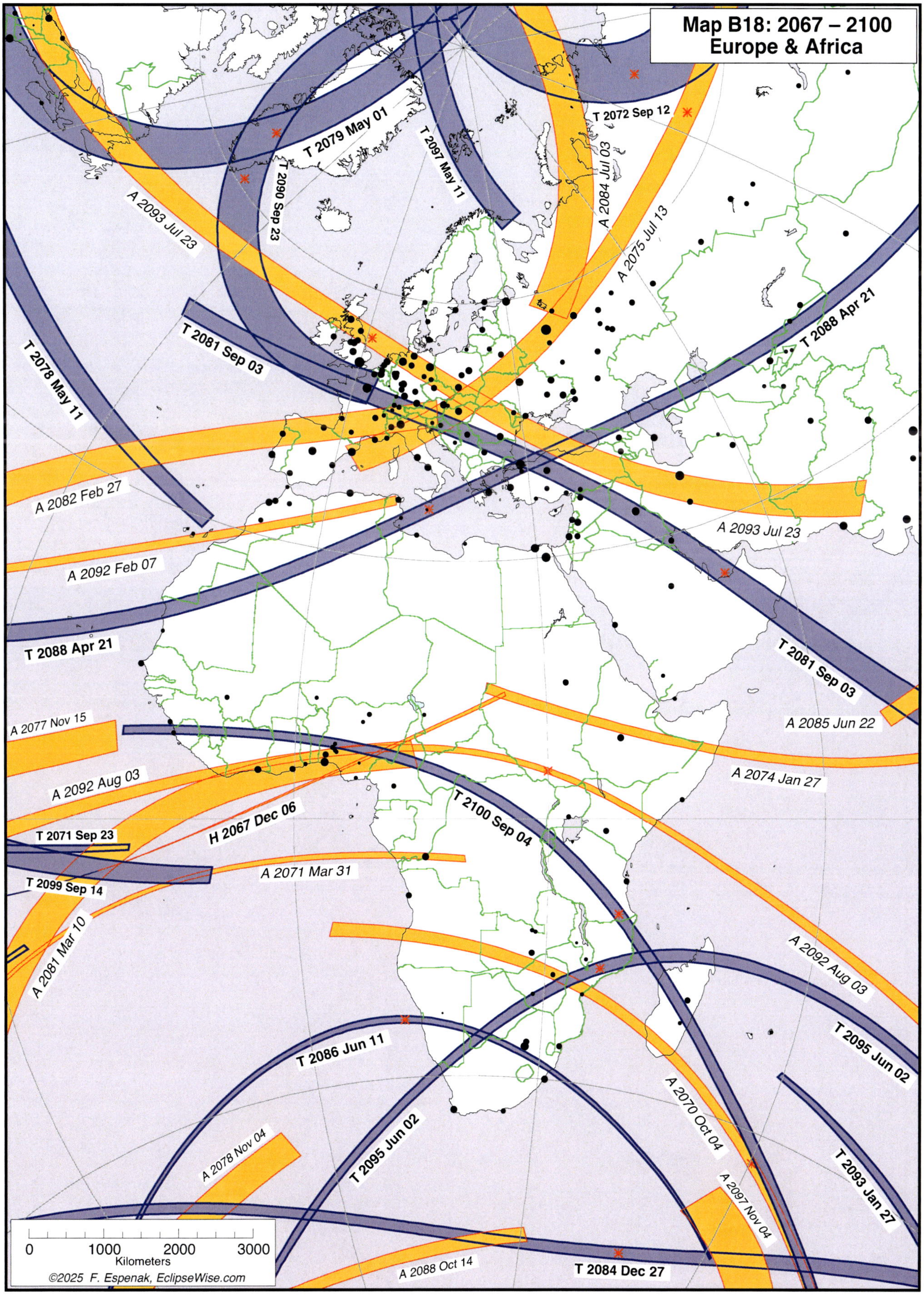
Map B18: 2067 – 2100
Europe & Africa
T 2079 May 01
T 2097 May 11
T 2072 Sep 12
T 2090 Sep 23
A 2084 Jul 03
A 2093 Jul 23
A 2075 Jul 13
T 2081 Sep 03
T 2088 Apr 21
T 2078 May 11
A 2082 Feb 27
A 2092 Feb 07
A 2093 Jul 23
T 2088 Apr 21
T 2081 Sep 03
A 2085 Jun 22
A 2077 Nov 15
A 2074 Jan 27
A 2092 Aug 03
H 2067 Dec 06
T 2100 Sep 04
T 2071 Sep 23
A 2071 Mar 31
T 2099 Sep 14
A 2081 Mar 10
A 2092 Aug 03
T 2086 Jun 11
T 2095 Jun 02
A 2070 Oct 04
T 2095 Jun 02
A 2078 Nov 04
A 2097 Nov 04
T 2093 Jan 27
A 2088 Oct 14
T 2084 Dec 27
0
1000
2000
3000
Kilometers
©2025 F. Espenak, EclipseWise.com

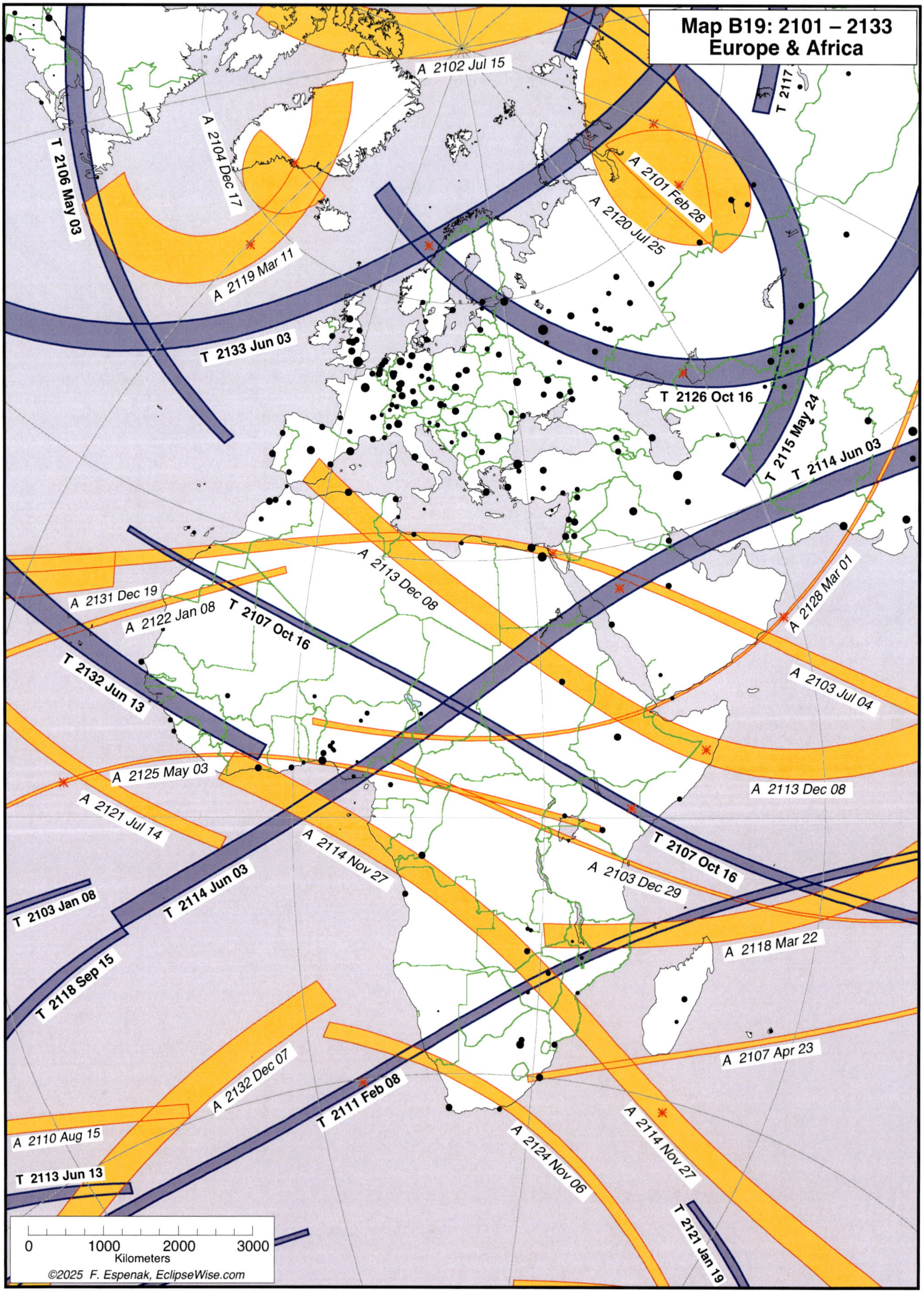
Map B19: 2101 – 2133
Europe & Africa
A 2102 Jul 15
T 2117
A 2104 Dec 17
T 2106 May 03
A 2101 Feb 28
A 2120 Jul 25
A 2119 Mar 11
T 2133 Jun 03
T 2126 Oct 16
T 2115 May 24
T 2114 Jun 03
A 2113 Dec 08
A 2131 Dec 19
A 2122 Jan 08
T 2107 Oct 16
A 2128 Mar 01
A 2103 Jul 04
T 2132 Jun 13
A 2125 May 03
A 2113 Dec 08
A 2121 Jul 14
A 2114 Nov 27
T 2107 Oct 16
A 2103 Dec 29
T 2103 Jan 08
T 2114 Jun 03
A 2118 Mar 22
T 2118 Sep 15
A 2107 Apr 23
A 2132 Dec 07
T 2111 Feb 08
A 2110 Aug 15
A 2114 Nov 27
A 2124 Nov 06
T 2113 Jun 13
T 2121 Jan 19
0 1000 2000 3000
Kilometers
©2025 F. Espenak, EclipseWise.com

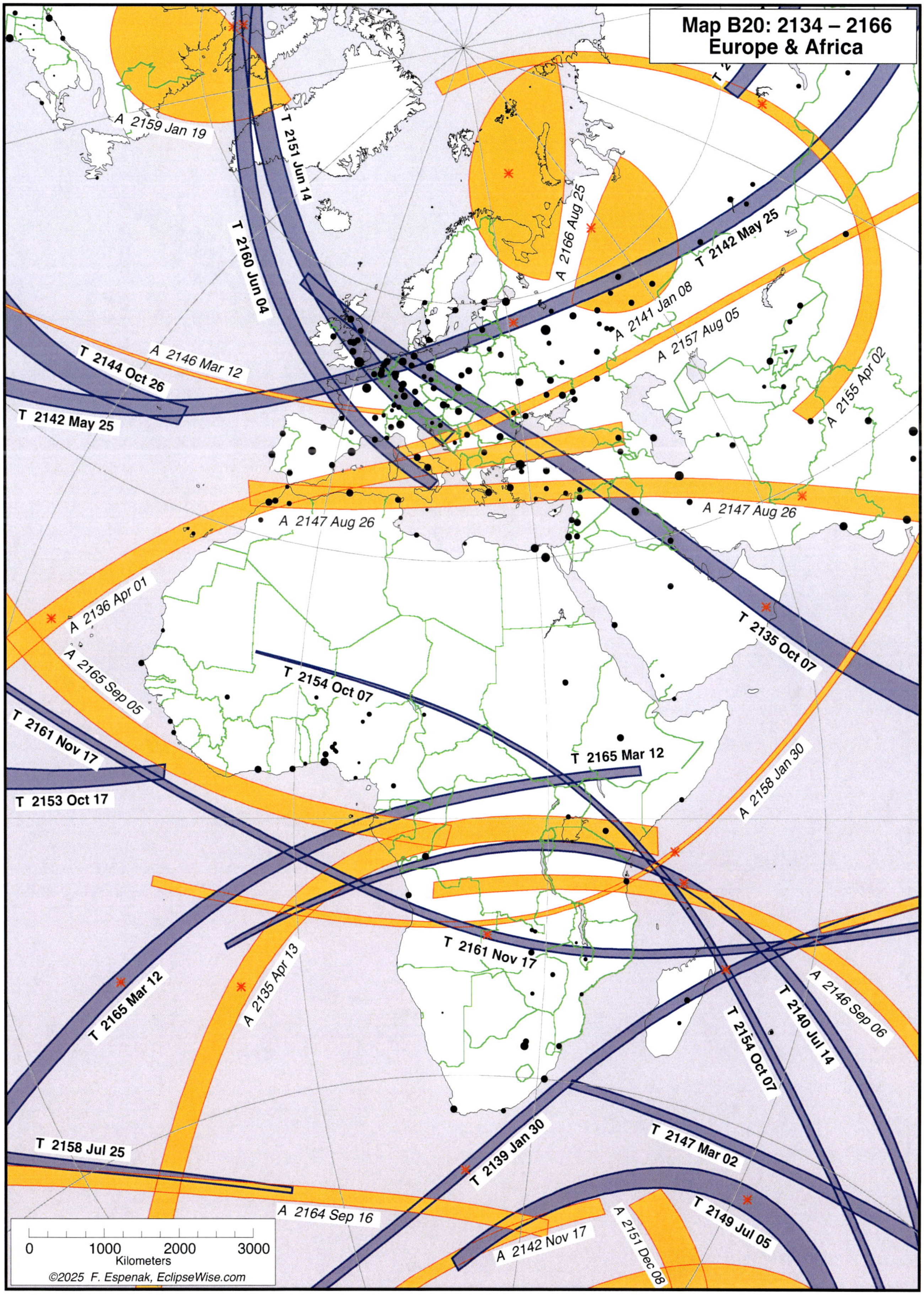
Map B20: 2134 – 2166
Europe & Africa
A 2159 Jan 19
T 2151 Jun 14
T 2160 Jun 04
A 2166 Aug 25
T 2142 May 25
A 2141 Jan 08
A 2157 Aug 05
A 2155 Apr 02
T 2144 Oct 26
A 2146 Mar 12
T 2142 May 25
A 2147 Aug 26
A 2147 Aug 26
A 2136 Apr 01
A 2165 Sep 05
T 2135 Oct 07
T 2154 Oct 07
T 2161 Nov 17
T 2165 Mar 12
T 2153 Oct 17
A 2158 Jan 30
T 2161 Nov 17
T 2165 Mar 12
A 2135 Apr 13
A 2146 Sep 06
T 2140 Jul 14
T 2154 Oct 07
T 2147 Mar 02
T 2139 Jan 30
T 2158 Jul 25
T 2149 Jul 05
A 2164 Sep 16
A 2151 Dec 08
A 2142 Nov 17
0
1000
2000
3000
Kilometers
©2025 F. Espenak, EclipseWise.com

Map B21: 2167 – 2200
Europe & Africa
A 2175 Aug 16
A 2191 Apr 23
T 2178 Jun 16
T 2200 Apr 14
T 2196 Jun 26
A 2173 Apr 12
T 2187 Jul 06
T 2189 Nov 08
A 2195 Feb 10
H 2182 Apr 03
T 2180 Nov 17
A 2186 Jan 20
A 2176 Feb 10
T 2187 Jul 06
A 2190 May 04
T 2180 Nov 17
A 2168 Jan 10
A 2168 Jan 10
A 2179 Jun 05
T 2194 Aug 16
T 2197 Dec 09
A 2190 May 04
T 2168 Jul 05
T 2193 Mar 03
T 2186 Jul 16
A 2187 Jan 09
A 2189 May 15
T 2193 Mar 03
A 2200 Oct 09
H 2172 Oct 17
A 2196 Dec 19
A 2180 May 24
A 2189 May 15
A 2168 Dec 29
T 2184 Mar 12
Oct 18
0 1000 2000 3000
Kilometers
©2025 F. Espenak, EclipseWise.com

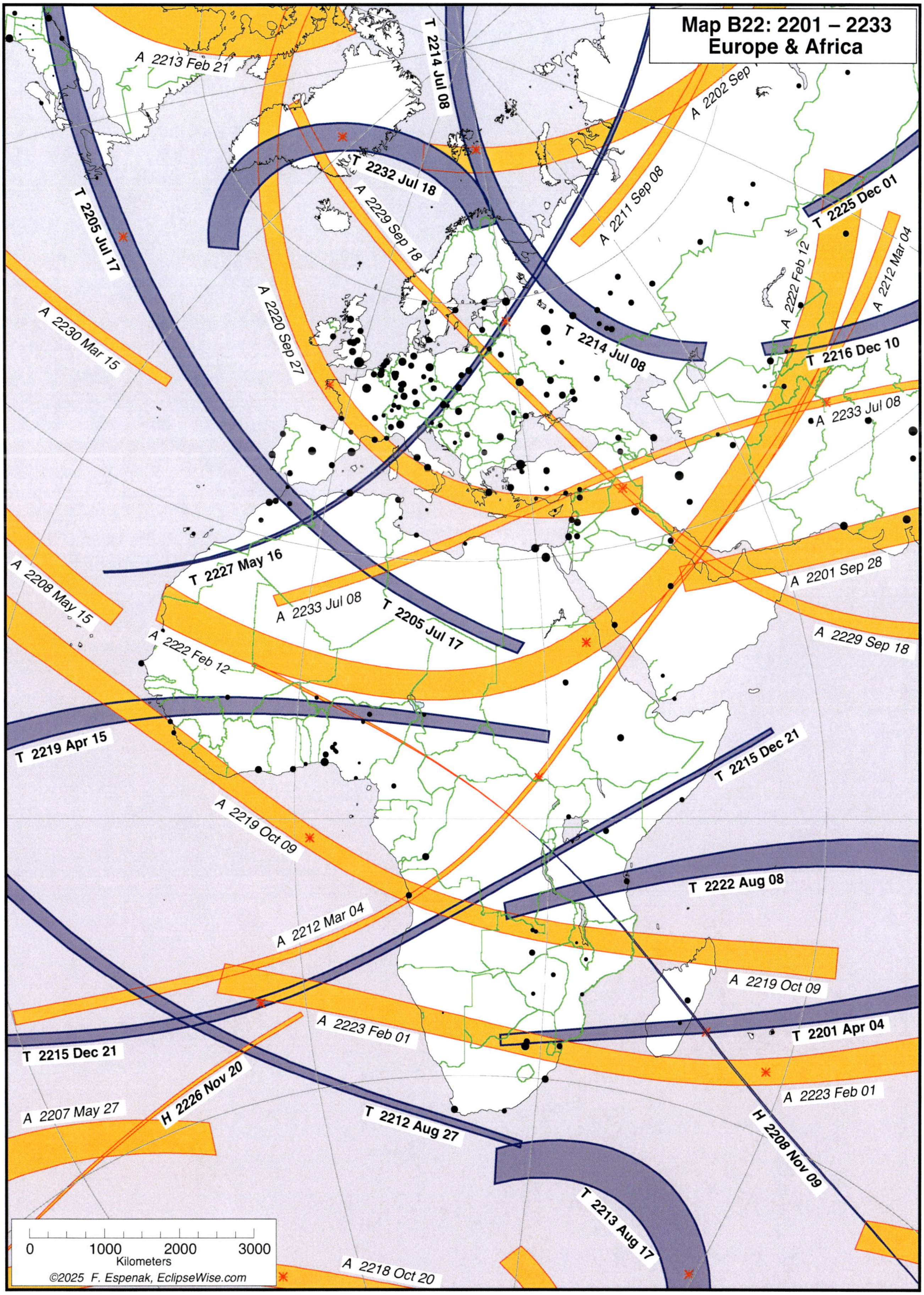
Map B22: 2201 – 2233
Europe & Africa
A 2213 Feb 21
T 2214 Jul 08
A 2202 Sep
T 2232 Jul 18
A 2229 Sep 18
T 2205 Jul 17
A 2211 Sep 08
T 2225 Dec 01
A 2212 Mar 04
A 2222 Feb 12
A 2220 Sep 27
A 2230 Mar 15
T 2214 Jul 08
T 2216 Dec 10
A 2233 Jul 08
T 2227 May 16
A 2201 Sep 28
A 2208 May 15
A 2233 Jul 08
T 2205 Jul 17
A 2229 Sep 18
A 2222 Feb 12
T 2219 Apr 15
T 2215 Dec 21
A 2219 Oct 09
T 2222 Aug 08
A 2212 Mar 04
A 2219 Oct 09
A 2223 Feb 01
T 2201 Apr 04
T 2215 Dec 21
H 2226 Nov 20
A 2223 Feb 01
A 2207 May 27
T 2212 Aug 27
H 2208 Nov 09
T 2213 Aug 17
0 1000 2000 3000
Kilometers
©2025 F. Espenak, EclipseWise.com
A 2218 Oct 20

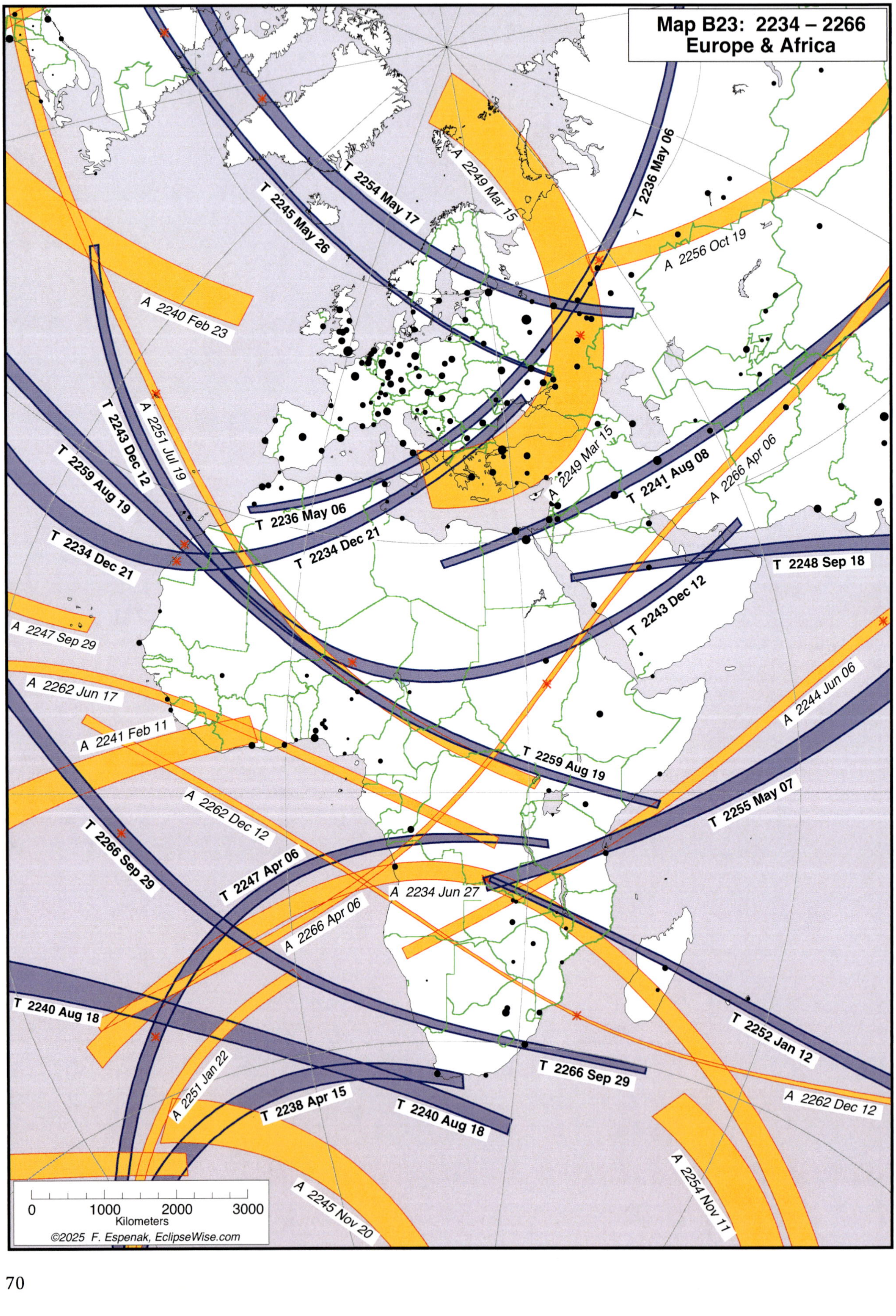
Map B23: 2234 – 2266
Europe & Africa
T 2254 May 17
T 2245 May 26
A 2249 Mar 15
T 2236 May 06
A 2256 Oct 19
A 2240 Feb 23
T 2243 Dec 12
A 2251 Jul 19
T 2259 Aug 19
T 2234 Dec 21
A 2249 Mar 15
T 2241 Aug 08
A 2266 Apr 06
T 2236 May 06
T 2234 Dec 21
T 2248 Sep 18
A 2247 Sep 29
T 2243 Dec 12
A 2262 Jun 17
A 2241 Feb 11
A 2244 Jun 06
T 2259 Aug 19
A 2262 Dec 12
T 2255 May 07
T 2266 Sep 29
T 2247 Apr 06
A 2234 Jun 27
A 2266 Apr 06
T 2240 Aug 18
T 2252 Jan 12
T 2266 Sep 29
A 2251 Jan 22
T 2238 Apr 15
T 2240 Aug 18
A 2262 Dec 12
A 2254 Nov 11
A 2245 Nov 20
0
1000
2000
3000
Kilometers
©2025 F. Espenak, EclipseWise.com

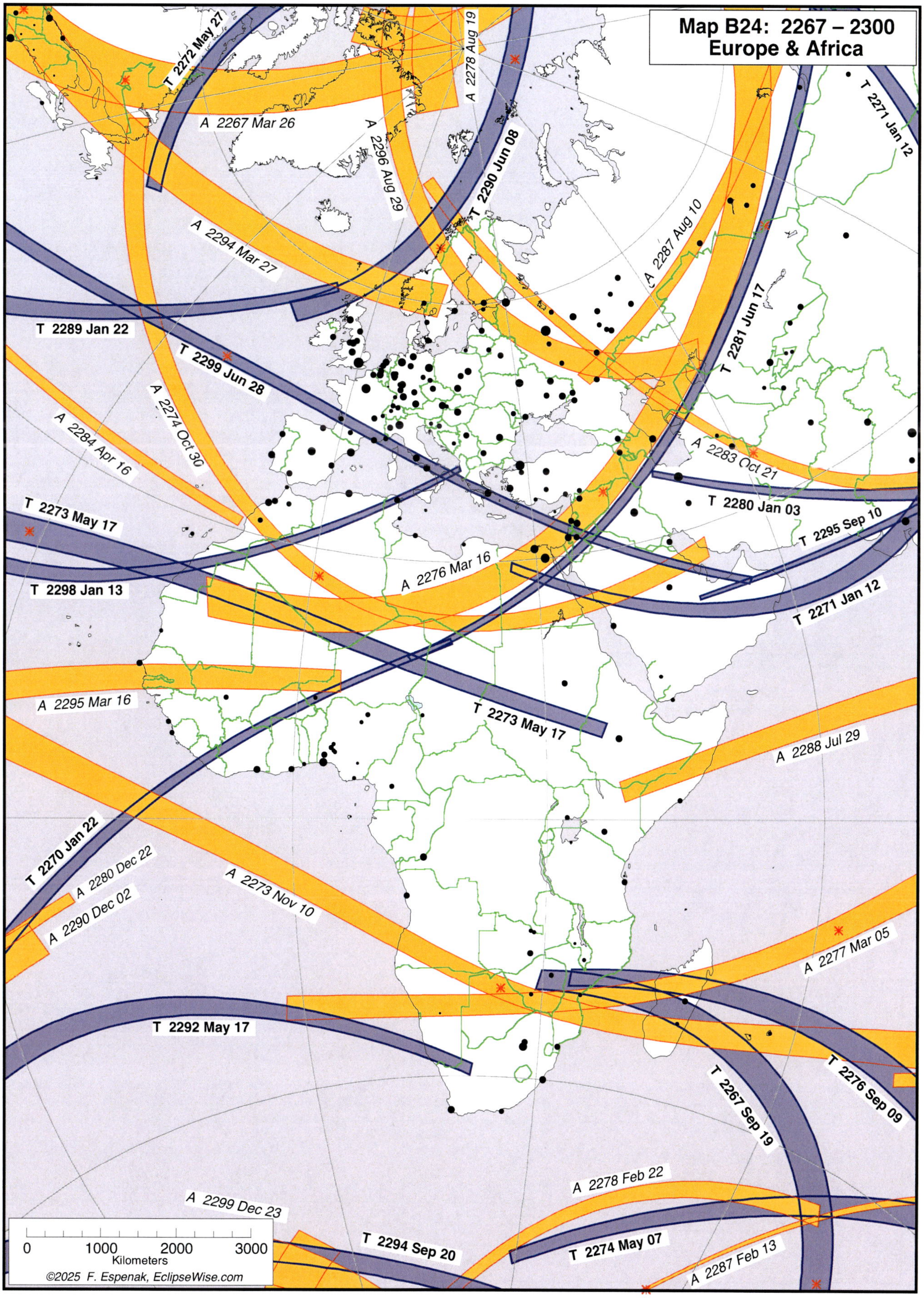
Map B24: 2267 – 2300
Europe & Africa
T 2272 May 27
A 2278 Aug 19
T 2271 Jan 12
A 2267 Mar 26
A 2296 Aug 29
T 2290 Jun 08
A 2294 Mar 27
A 2287 Aug 10
T 2289 Jan 22
T 2299 Jun 28
T 2281 Jun 17
A 2274 Oct 30
A 2284 Apr 16
A 2283 Oct 21
T 2280 Jan 03
T 2273 May 17
T 2295 Sep 10
A 2276 Mar 16
T 2298 Jan 13
T 2271 Jan 12
A 2295 Mar 16
T 2273 May 17
A 2288 Jul 29
T 2270 Jan 22
A 2280 Dec 22
A 2290 Dec 02
A 2273 Nov 10
A 2277 Mar 05
T 2292 May 17
T 2267 Sep 19
T 2276 Sep 09
A 2278 Feb 22
A 2299 Dec 23
T 2294 Sep 20
T 2274 May 07
A 2287 Feb 13
0 1000 2000 3000
Kilometers
©2025 F. Espenak, EclipseWise.com

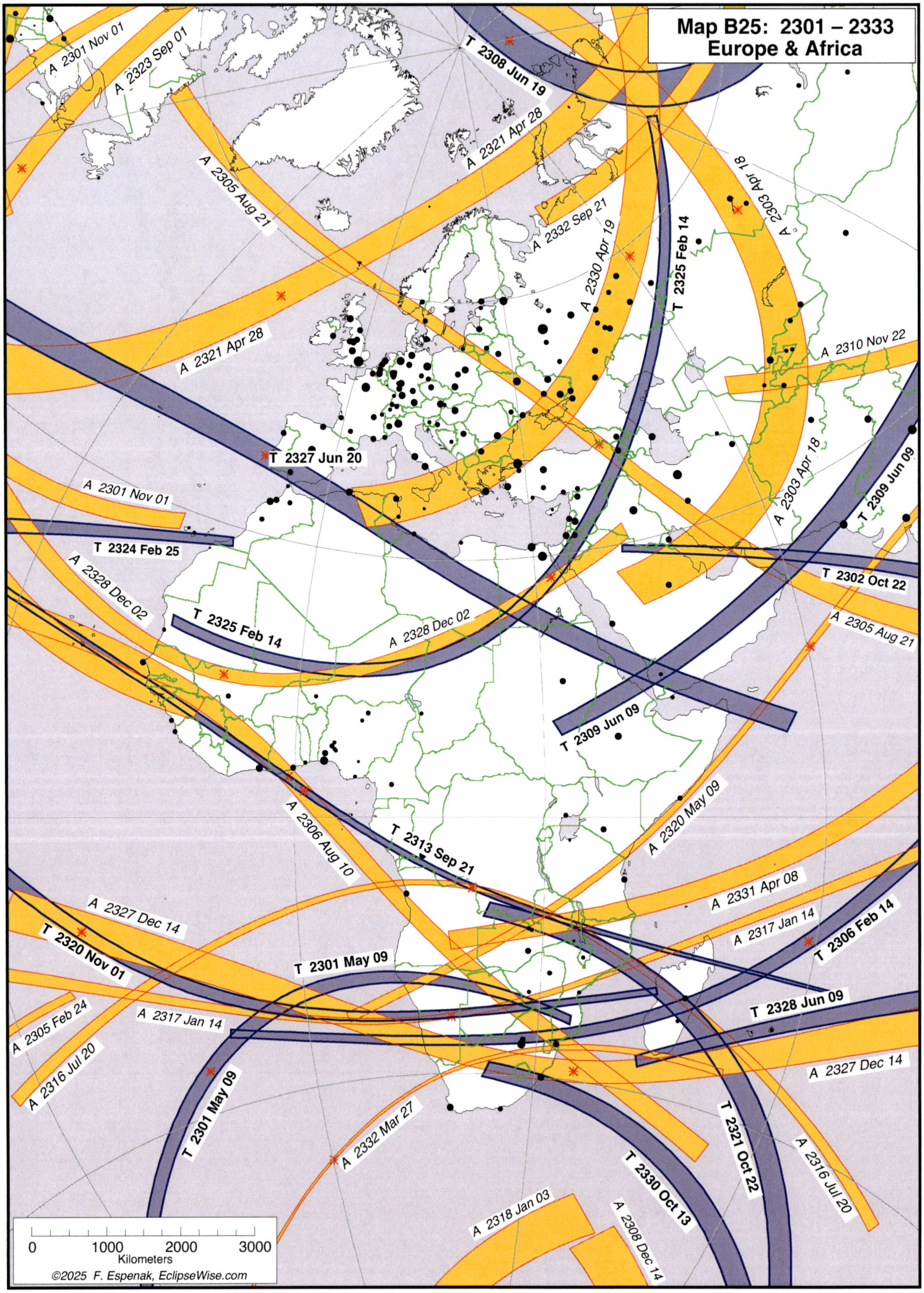

Map B25: 2301 – 2333
Europe & Africa
A 2301 Nov 01
A 2323 Sep 01
T 2308 Jun 19
A 2321 Apr 28
A 2305 Aug 21
A 2303 Apr 18
A 2332 Sep 21
A 2330 Apr 19
T 2325 Feb 14
A 2321 Apr 28
A 2310 Nov 22
T 2327 Jun 20
A 2301 Nov 01
A 2303 Apr 18
T 2309 Jun 09
T 2324 Feb 25
A 2328 Dec 02
T 2302 Oct 22
T 2325 Feb 14
A 2328 Dec 02
A 2305 Aug 21
T 2309 Jun 09
A 2320 May 09
A 2306 Aug 10
T 2313 Sep 21
A 2331 Apr 08
A 2327 Dec 14
A 2317 Jan 14
T 2306 Feb 14
T 2320 Nov 01
T 2301 May 09
T 2328 Jun 09
A 2305 Feb 24
A 2317 Jan 14
A 2316 Jul 20
A 2327 Dec 14
T 2301 May 09
A 2332 Mar 27
T 2321 Oct 22
A 2316 Jul 20
T 2330 Oct 13
A 2318 Jan 03
A 2308 Dec 14
0 1000 2000 3000
Kilometers
©2025 F. Espenak, EclipseWise.com

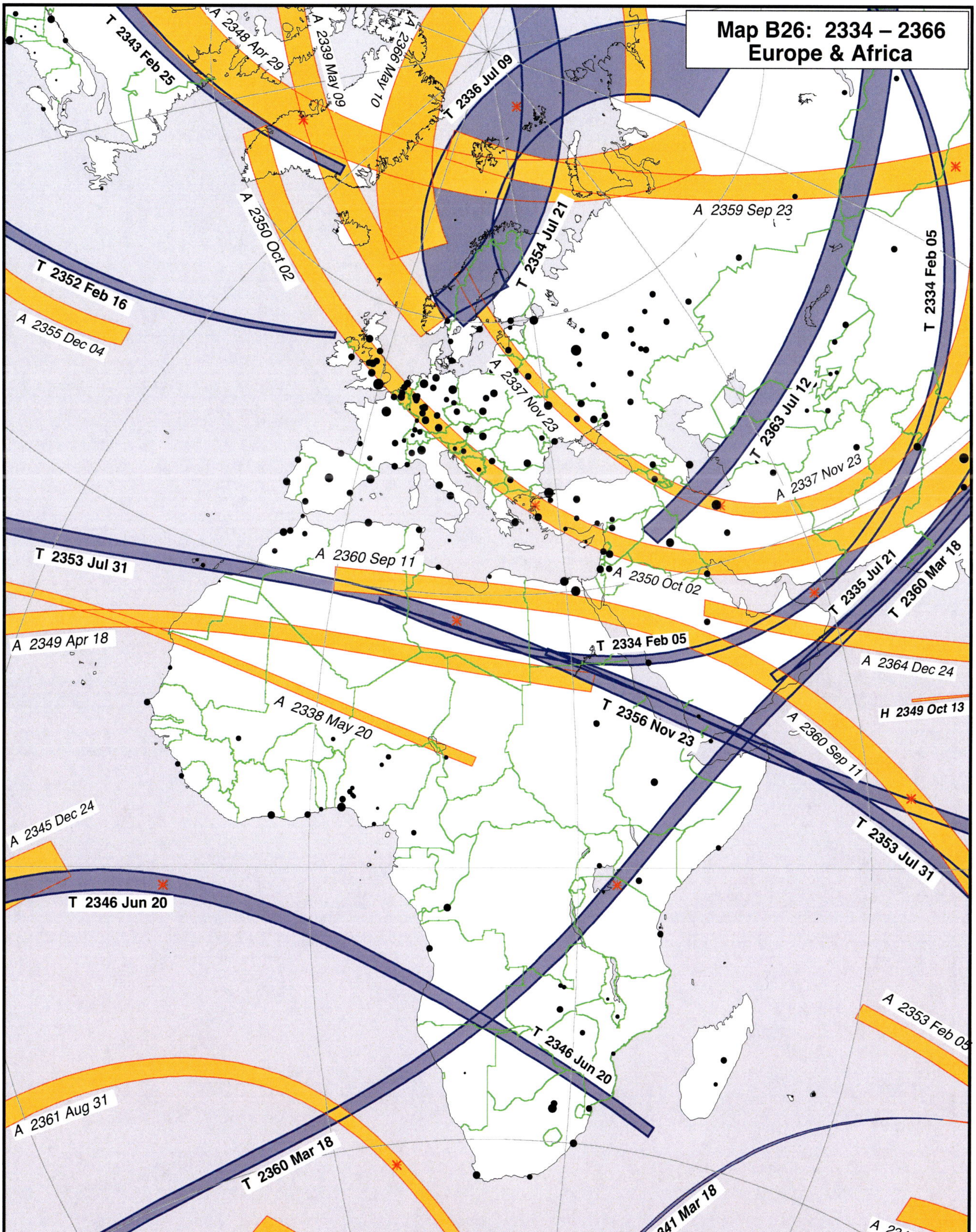
Map B26: 2334 – 2366
Europe & Africa
T 2343 Feb 25
A 2348 Apr 29
A 2339 May 09
A 2366 May 10
T 2336 Jul 09
A 2359 Sep 23
A 2350 Oct 02
T 2354 Jul 21
T 2334 Feb 05
T 2352 Feb 16
A 2355 Dec 04
A 2337 Nov 23
T 2363 Jul 12
A 2337 Nov 23
T 2353 Jul 31
A 2360 Sep 11
A 2350 Oct 02
T 2335 Jul 21
T 2360 Mar 18
A 2349 Apr 18
T 2334 Feb 05
A 2364 Dec 24
H 2349 Oct 13
A 2338 May 20
T 2356 Nov 23
A 2360 Sep 11
A 2345 Dec 24
T 2353 Jul 31
T 2346 Jun 20
A 2353 Feb 05
T 2346 Jun 20
A 2361 Aug 31
T 2360 Mar 18
H 2341 Mar 18
A 2343 Aug 21
Jan 04
T 2355 Jun 11
A 2361 Aug 31
0 1000 2000 3000
Kilometers
©2025 F. Espenak, EclipseWise.com

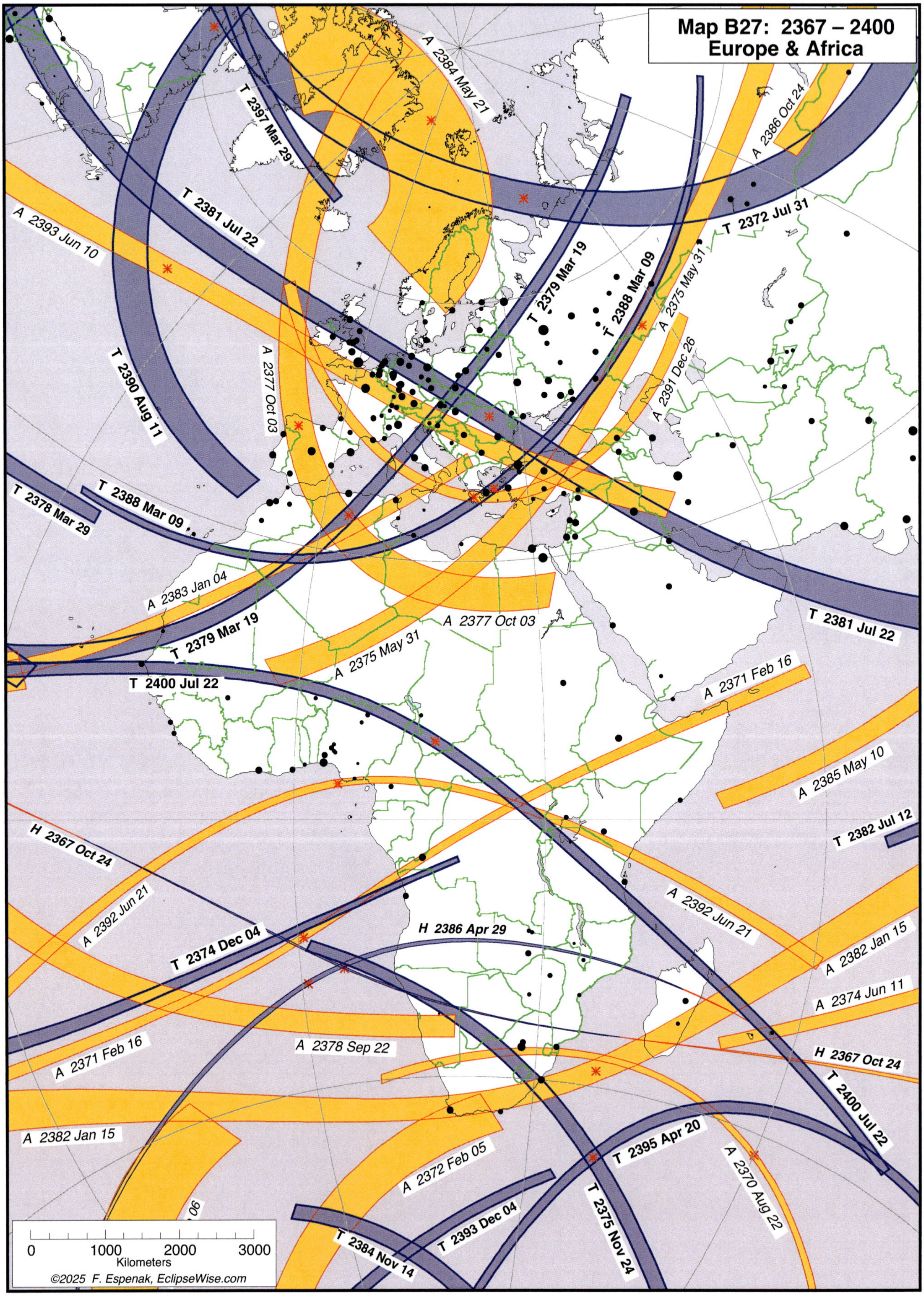
Map B27: 2367 – 2400
Europe & Africa
A 2384 May 21
T 2397 Mar 29
A 2386 Oct 24
T 2381 Jul 22
A 2393 Jun 10
T 2372 Jul 31
T 2379 Mar 19
T 2388 Mar 09
A 2375 May 31
A 2391 Dec 26
T 2390 Aug 11
A 2377 Oct 03
T 2378 Mar 29
T 2388 Mar 09
A 2383 Jan 04
A 2377 Oct 03
T 2381 Jul 22
T 2379 Mar 19
A 2375 May 31
T 2400 Jul 22
A 2371 Feb 16
A 2385 May 10
T 2382 Jul 12
H 2367 Oct 24
A 2392 Jun 21
T 2374 Dec 04
H 2386 Apr 29
A 2392 Jun 21
A 2382 Jan 15
A 2374 Jun 11
A 2378 Sep 22
A 2371 Feb 16
H 2367 Oct 24
T 2400 Jul 22
A 2382 Jan 15
A 2372 Feb 05
T 2395 Apr 20
A 2370 Aug 22
T 2393 Dec 04
T 2375 Nov 24
T 2384 Nov 14
0 1000 2000 3000
Kilometers
©2025 F. Espenak, EclipseWise.com

Appendix C

World Atlas of Central Solar Eclipses

Asia & Australia: 1501 to 2400

Index for Appendix C

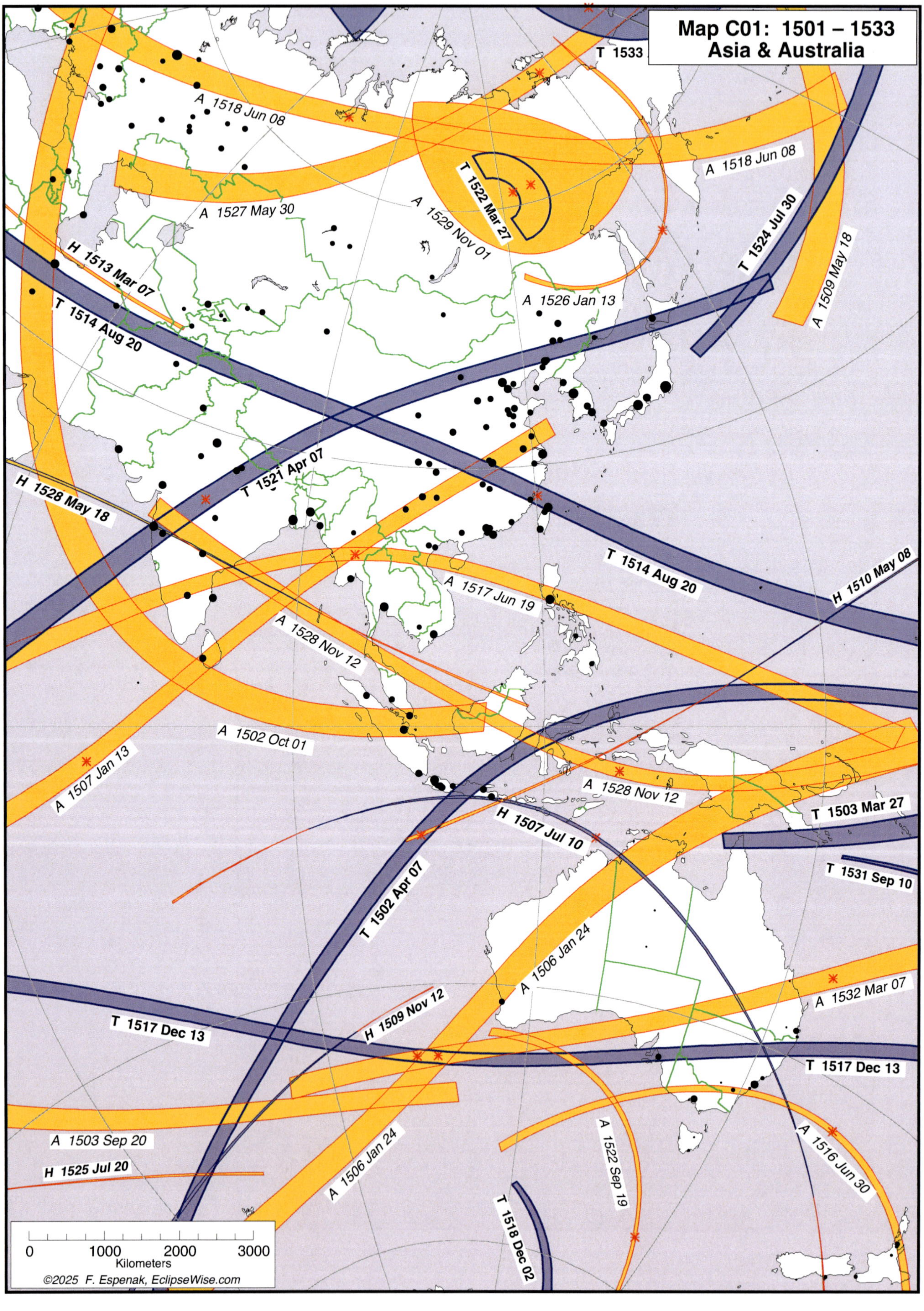
Map C01: 1501 – 1533
Asia & Australia
T 1533
A 1518 Jun 08
A 1518 Jun 08
A 1527 May 30
T 1522 Mar 27
A 1529 Nov 01
T 1524 Jul 30
A 1509 May 18
H 1513 Mar 07
T 1514 Aug 20
A 1526 Jan 13
T 1521 Apr 07
H 1528 May 18
T 1514 Aug 20
H 1510 May 08
A 1517 Jun 19
A 1528 Nov 12
A 1502 Oct 01
A 1507 Jan 13
A 1528 Nov 12
T 1503 Mar 27
H 1507 Jul 10
T 1531 Sep 10
T 1502 Apr 07
A 1506 Jan 24
A 1532 Mar 07
H 1509 Nov 12
T 1517 Dec 13
T 1517 Dec 13
A 1503 Sep 20
A 1522 Sep 19
A 1516 Jun 30
A 1506 Jan 24
H 1525 Jul 20
T 1518 Dec 02
0 1000 2000 3000
Kilometers
©2025 F. Espenak, EclipseWise.com

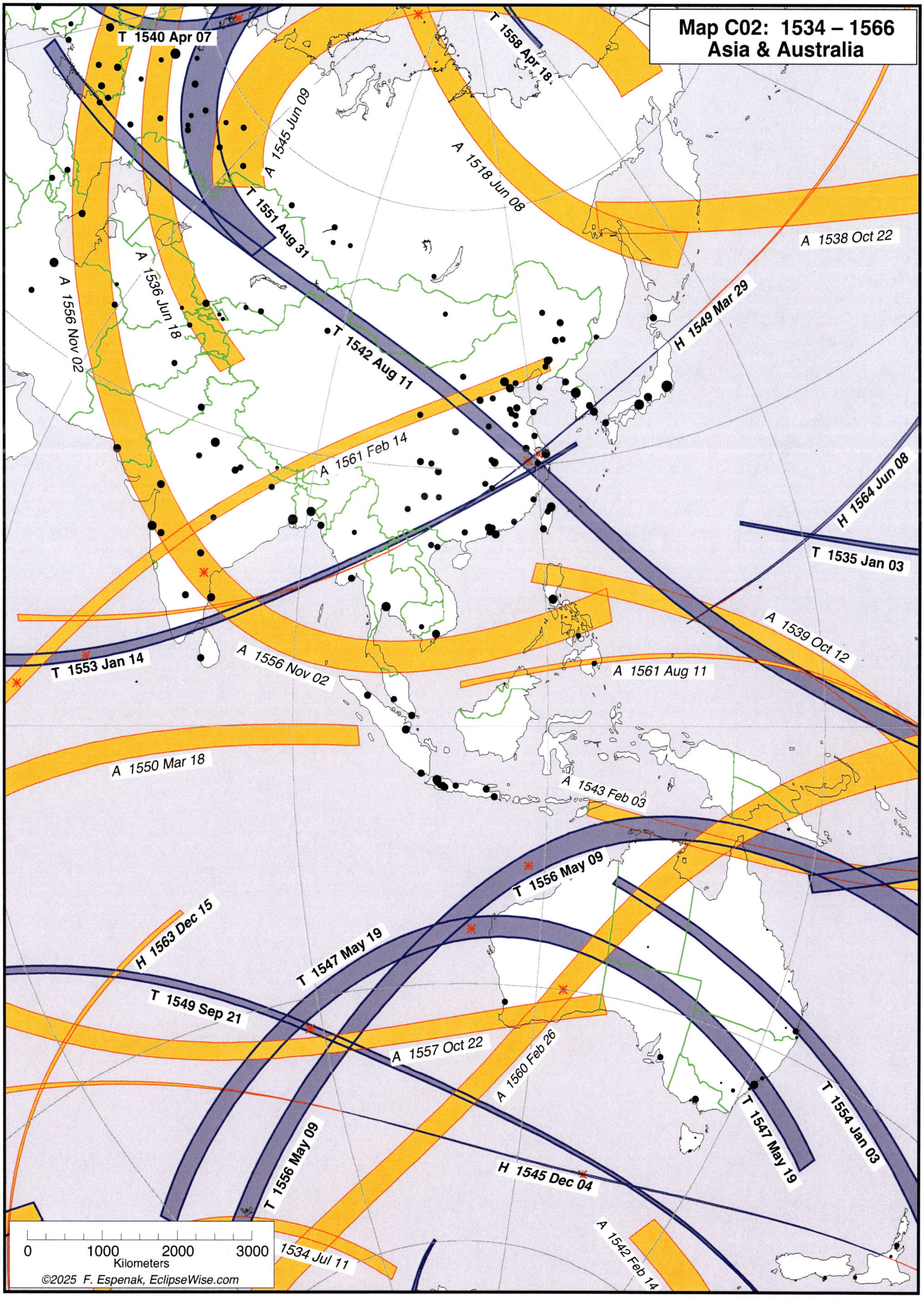
Map C02: 1534 – 1566
Asia & Australia
T 1540 Apr 07
T 1558 Apr 18
A 1545 Jun 09
A 1518 Jun 08
T 1551 Aug 31
A 1538 Oct 22
A 1536 Jun 18
A 1556 Nov 02
H 1549 Mar 29
T 1542 Aug 11
A 1561 Feb 14
H 1564 Jun 08
T 1535 Jan 03
A 1539 Oct 12
T 1553 Jan 14
A 1556 Nov 02
A 1561 Aug 11
A 1550 Mar 18
A 1543 Feb 03
T 1556 May 09
H 1563 Dec 15
T 1547 May 19
T 1549 Sep 21
A 1557 Oct 22
A 1560 Feb 26
T 1554 Jan 03
T 1547 May 19
T 1556 May 09
H 1545 Dec 04
A 1542 Feb 14
1534 Jul 11
0 1000 2000 3000
Kilometers
©2025 F. Espenak, EclipseWise.com

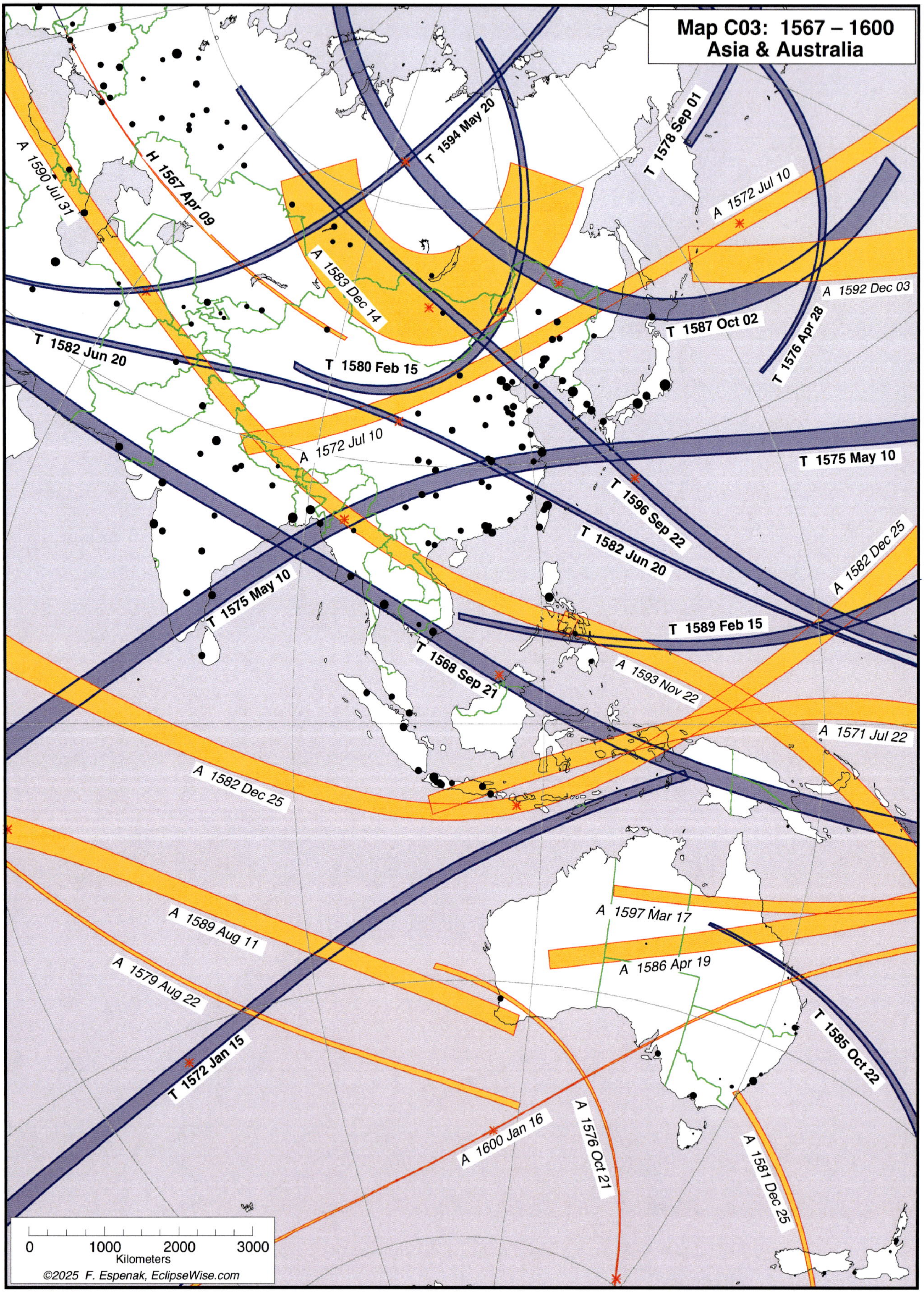
Map C03: 1567 – 1600
Asia & Australia
A 1590 Jul 31
H 1567 Apr 09
T 1594 May 20
T 1578 Sep 01
A 1572 Jul 10
A 1583 Dec 14
A 1592 Dec 03
T 1587 Oct 02
T 1576 Apr 28
T 1582 Jun 20
T 1580 Feb 15
A 1572 Jul 10
T 1575 May 10
T 1596 Sep 22
T 1582 Jun 20
A 1582 Dec 25
T 1575 May 10
T 1589 Feb 15
T 1568 Sep 21
A 1593 Nov 22
A 1571 Jul 22
A 1582 Dec 25
A 1597 Mar 17
A 1589 Aug 11
A 1586 Apr 19
A 1579 Aug 22
T 1585 Oct 22
T 1572 Jan 15
A 1600 Jan 16
A 1576 Oct 21
A 1581 Dec 25
0 1000 2000 3000
Kilometers
©2025 F. Espenak, EclipseWise.com

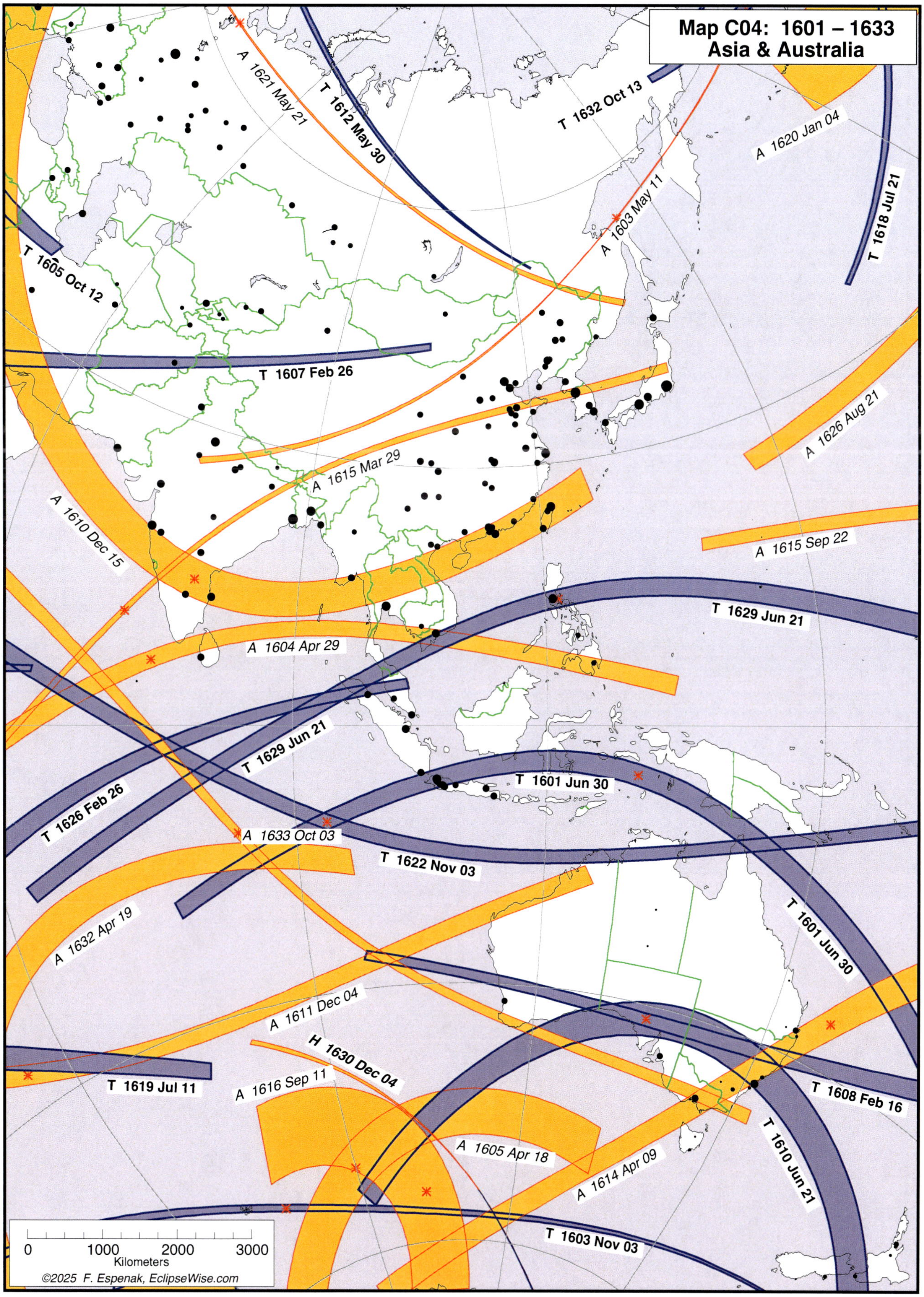
Map C04: 1601 – 1633
Asia & Australia
A 1621 May 21
T 1612 May 30
T 1632 Oct 13
A 1620 Jan 04
T 1618 Jul 21
A 1603 May 11
T 1605 Oct 12
T 1607 Feb 26
A 1626 Aug 21
A 1615 Mar 29
A 1610 Dec 15
A 1615 Sep 22
T 1629 Jun 21
A 1604 Apr 29
T 1629 Jun 21
T 1626 Feb 26
T 1601 Jun 30
A 1633 Oct 03
T 1622 Nov 03
A 1632 Apr 19
T 1601 Jun 30
A 1611 Dec 04
H 1630 Dec 04
A 1616 Sep 11
T 1619 Jul 11
T 1608 Feb 16
A 1605 Apr 18
A 1614 Apr 09
T 1610 Jun 21
T 1603 Nov 03
0
1000
2000
3000
Kilometers
©2025 F. Espenak, EclipseWise.com

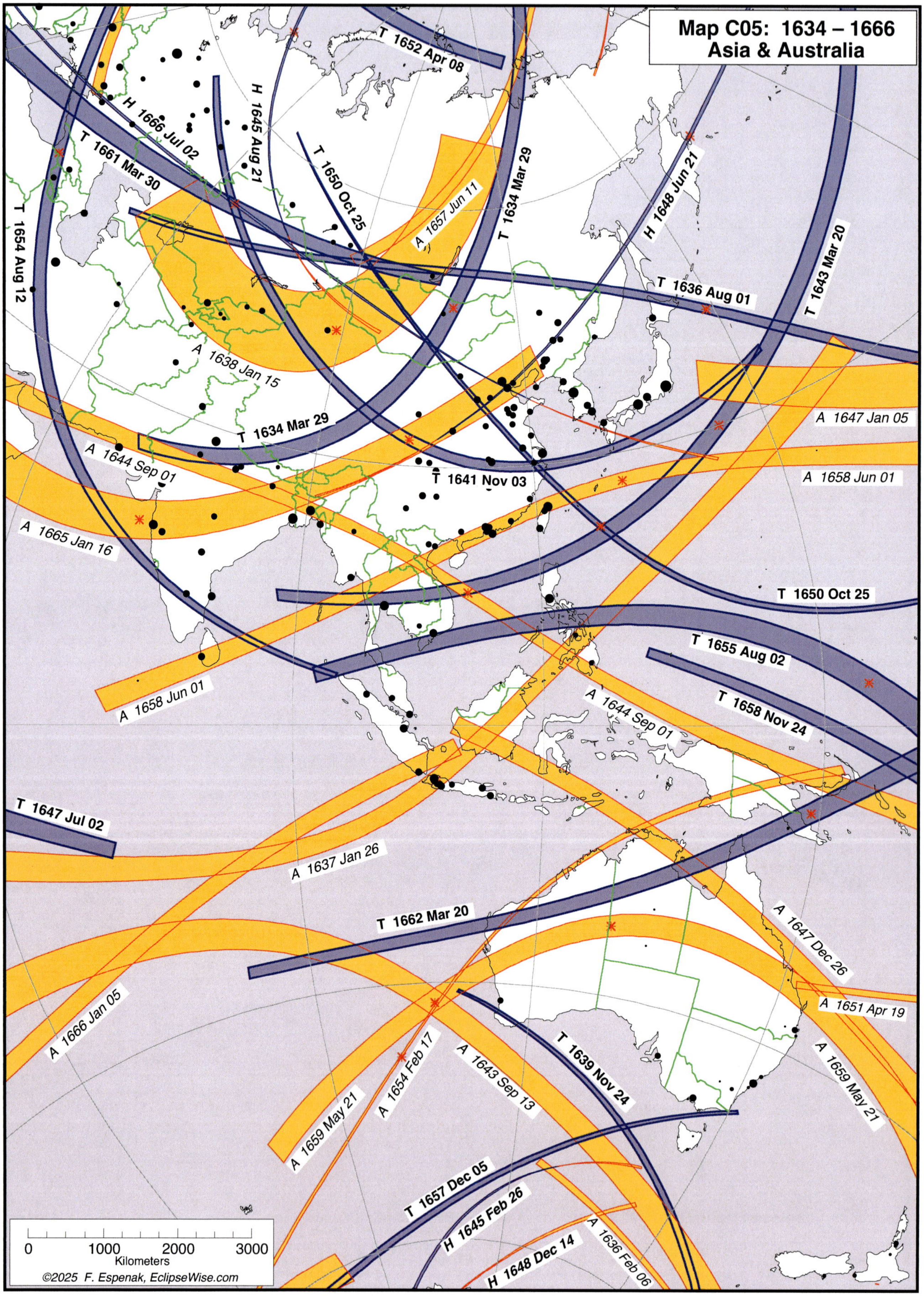
Map C05: 1634 – 1666
Asia & Australia
T 1652 Apr 08
H 1666 Jul 02
H 1645 Aug 21
T 1661 Mar 30
T 1650 Oct 25
A 1657 Jun 11
T 1634 Mar 29
H 1648 Jun 21
T 1654 Aug 12
T 1636 Aug 01
T 1643 Mar 20
A 1638 Jan 15
T 1634 Mar 29
A 1647 Jan 05
A 1644 Sep 01
T 1641 Nov 03
A 1658 Jun 01
A 1665 Jan 16
T 1650 Oct 25
T 1655 Aug 02
A 1658 Jun 01
A 1644 Sep 01
T 1658 Nov 24
T 1647 Jul 02
A 1637 Jan 26
T 1662 Mar 20
A 1647 Dec 26
A 1651 Apr 19
A 1666 Jan 05
A 1654 Feb 17
A 1643 Sep 13
T 1639 Nov 24
A 1659 May 21
A 1659 May 21
T 1657 Dec 05
H 1645 Feb 26
H 1648 Dec 14
A 1636 Feb 06
0 1000 2000 3000
Kilometers
©2025 F. Espenak, EclipseWise.com

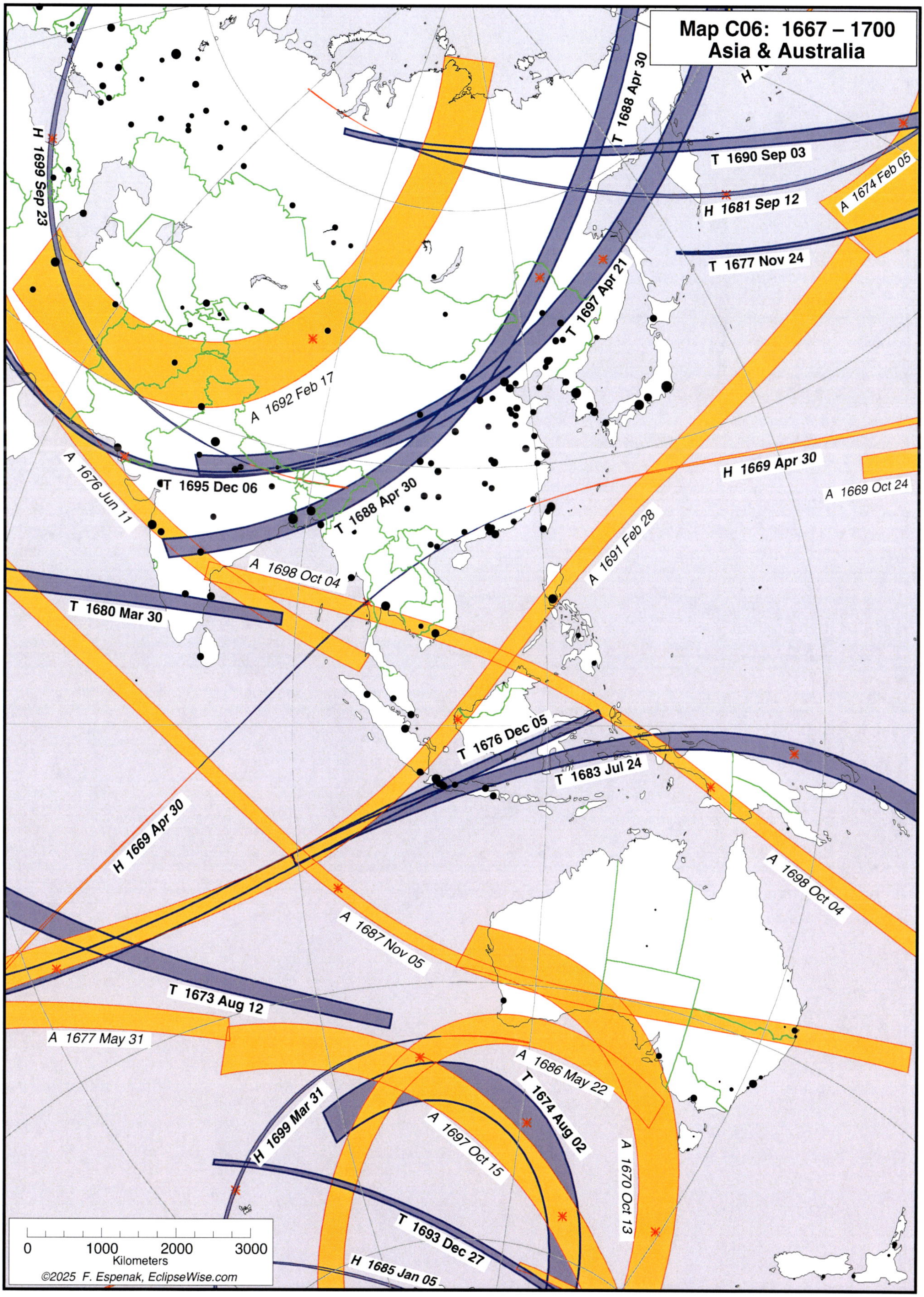
Map C06: 1667 – 1700
Asia & Australia
H 1699 Sep 23
T 1688 Apr 30
T 1690 Sep 03
H 1681 Sep 12
A 1674 Feb 05
T 1677 Nov 24
T 1697 Apr 21
A 1692 Feb 17
A 1676 Jun 11
T 1695 Dec 06
T 1688 Apr 30
H 1669 Apr 30
A 1669 Oct 24
A 1691 Feb 28
A 1698 Oct 04
T 1680 Mar 30
T 1676 Dec 05
T 1683 Jul 24
H 1669 Apr 30
A 1698 Oct 04
A 1687 Nov 05
T 1673 Aug 12
A 1677 May 31
A 1686 May 22
T 1674 Aug 02
H 1699 Mar 31
A 1697 Oct 15
A 1670 Oct 13
T 1693 Dec 27
H 1685 Jan 05
0 1000 2000 3000
Kilometers
©2025 F. Espenak, EclipseWise.com

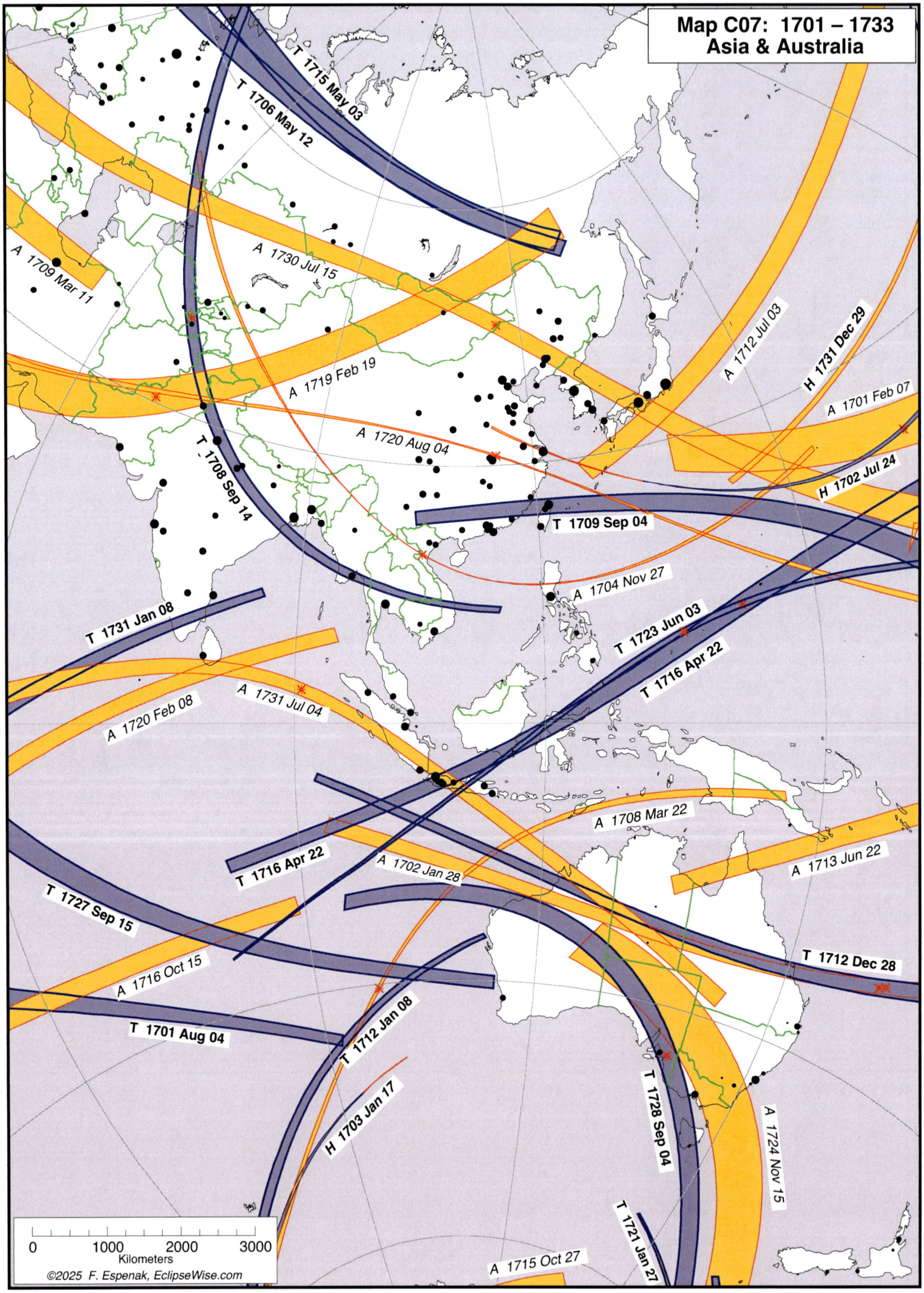
Map C07: 1701 – 1733
Asia & Australia
T 1715 May 03
T 1706 May 12
A 1709 Mar 11
A 1730 Jul 15
A 1712 Jul 03
H 1731 Dec 29
A 1719 Feb 19
A 1701 Feb 07
A 1720 Aug 04
T 1708 Sep 14
H 1702 Jul 24
T 1709 Sep 04
A 1704 Nov 27
T 1731 Jan 08
T 1723 Jun 03
T 1716 Apr 22
A 1731 Jul 04
A 1720 Feb 08
A 1708 Mar 22
T 1716 Apr 22
A 1702 Jan 28
A 1713 Jun 22
T 1727 Sep 15
T 1712 Dec 28
A 1716 Oct 15
T 1701 Aug 04
T 1712 Jan 08
H 1703 Jan 17
T 1728 Sep 04
A 1724 Nov 15
T 1721 Jan 27
A 1715 Oct 27
0 1000 2000 3000
Kilometers
©2025 F. Espenak, EclipseWise.com

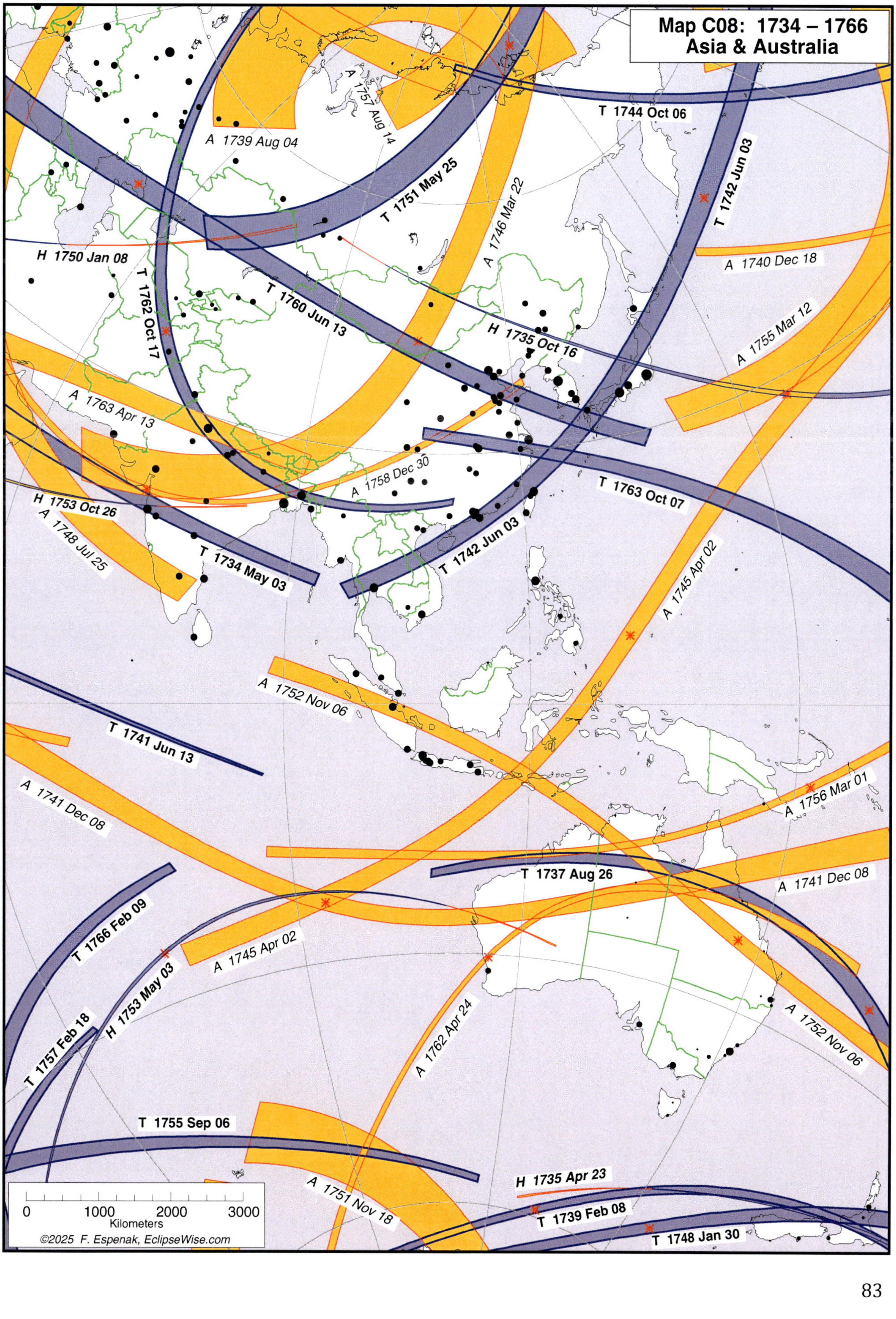
Map C08: 1734 – 1766
Asia & Australia
A 1757 Aug 14
A 1739 Aug 04
T 1744 Oct 06
T 1751 May 25
A 1746 Mar 22
T 1742 Jun 03
H 1750 Jan 08
A 1740 Dec 18
T 1762 Oct 17
T 1760 Jun 13
H 1735 Oct 16
A 1755 Mar 12
A 1763 Apr 13
A 1758 Dec 30
T 1763 Oct 07
H 1753 Oct 26
A 1748 Jul 25
T 1734 May 03
T 1742 Jun 03
A 1745 Apr 02
A 1752 Nov 06
T 1741 Jun 13
A 1741 Dec 08
A 1756 Mar 01
T 1737 Aug 26
A 1741 Dec 08
T 1766 Feb 09
A 1745 Apr 02
H 1753 May 03
A 1762 Apr 24
A 1752 Nov 06
T 1757 Feb 18
T 1755 Sep 06
A 1751 Nov 18
H 1735 Apr 23
T 1739 Feb 08
T 1748 Jan 30
0 1000 2000 3000
Kilometers
©2025 F. Espenak, EclipseWise.com

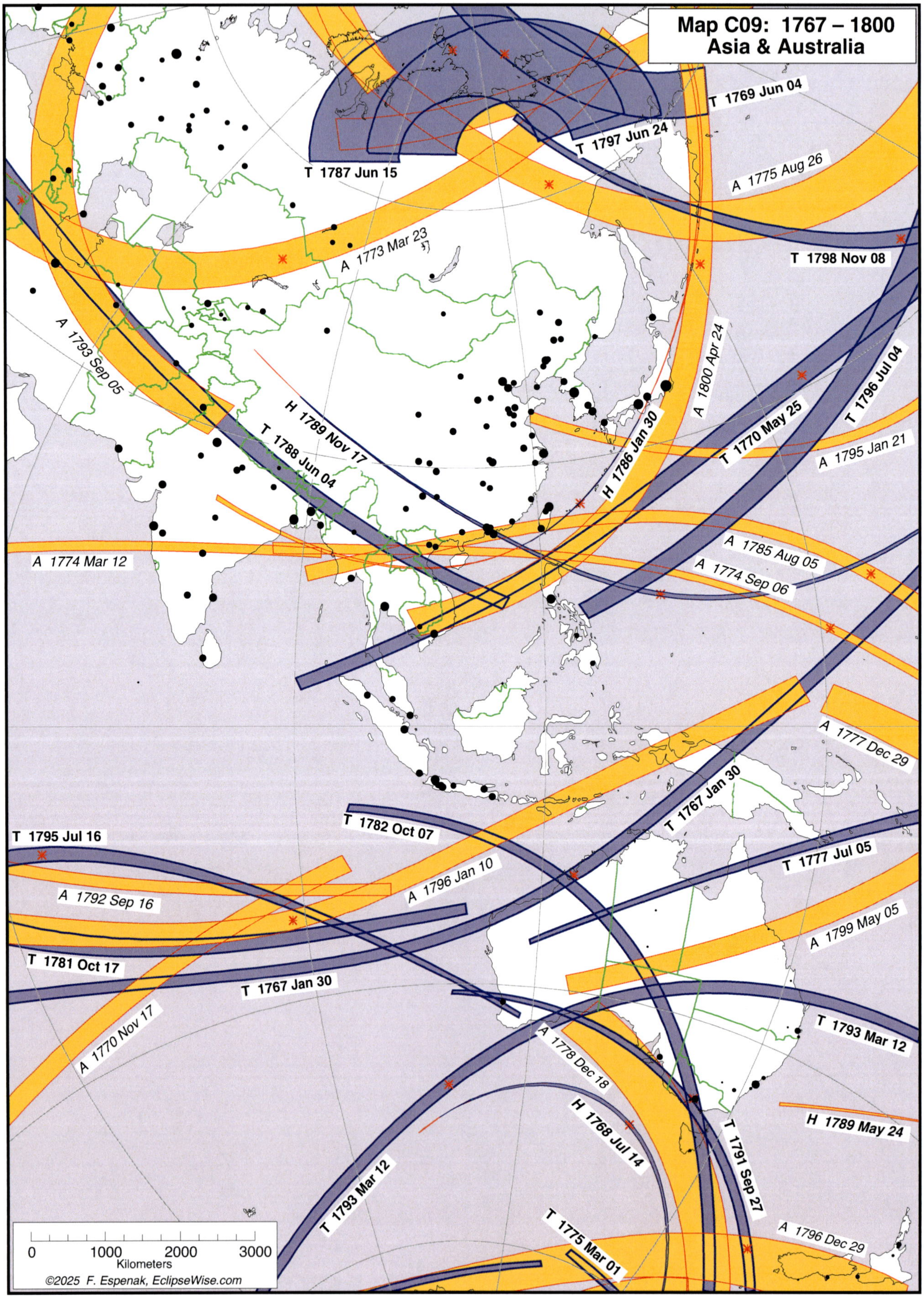
Map C09: 1767 – 1800
Asia & Australia
T 1769 Jun 04
T 1797 Jun 24
T 1787 Jun 15
A 1775 Aug 26
T 1798 Nov 08
A 1773 Mar 23
A 1793 Sep 05
A 1800 Apr 24
T 1796 Jul 04
H 1789 Nov 17
T 1788 Jun 04
H 1786 Jan 30
T 1770 May 25
A 1795 Jan 21
A 1774 Mar 12
A 1785 Aug 05
A 1774 Sep 06
A 1777 Dec 29
T 1767 Jan 30
T 1782 Oct 07
T 1795 Jul 16
T 1777 Jul 05
A 1796 Jan 10
A 1792 Sep 16
A 1799 May 05
T 1781 Oct 17
T 1767 Jan 30
A 1770 Nov 17
T 1793 Mar 12
A 1778 Dec 18
H 1768 Jul 14
H 1789 May 24
T 1791 Sep 27
T 1793 Mar 12
T 1775 Mar 01
A 1796 Dec 29
0 1000 2000 3000
Kilometers
©2025 F. Espenak, EclipseWise.com

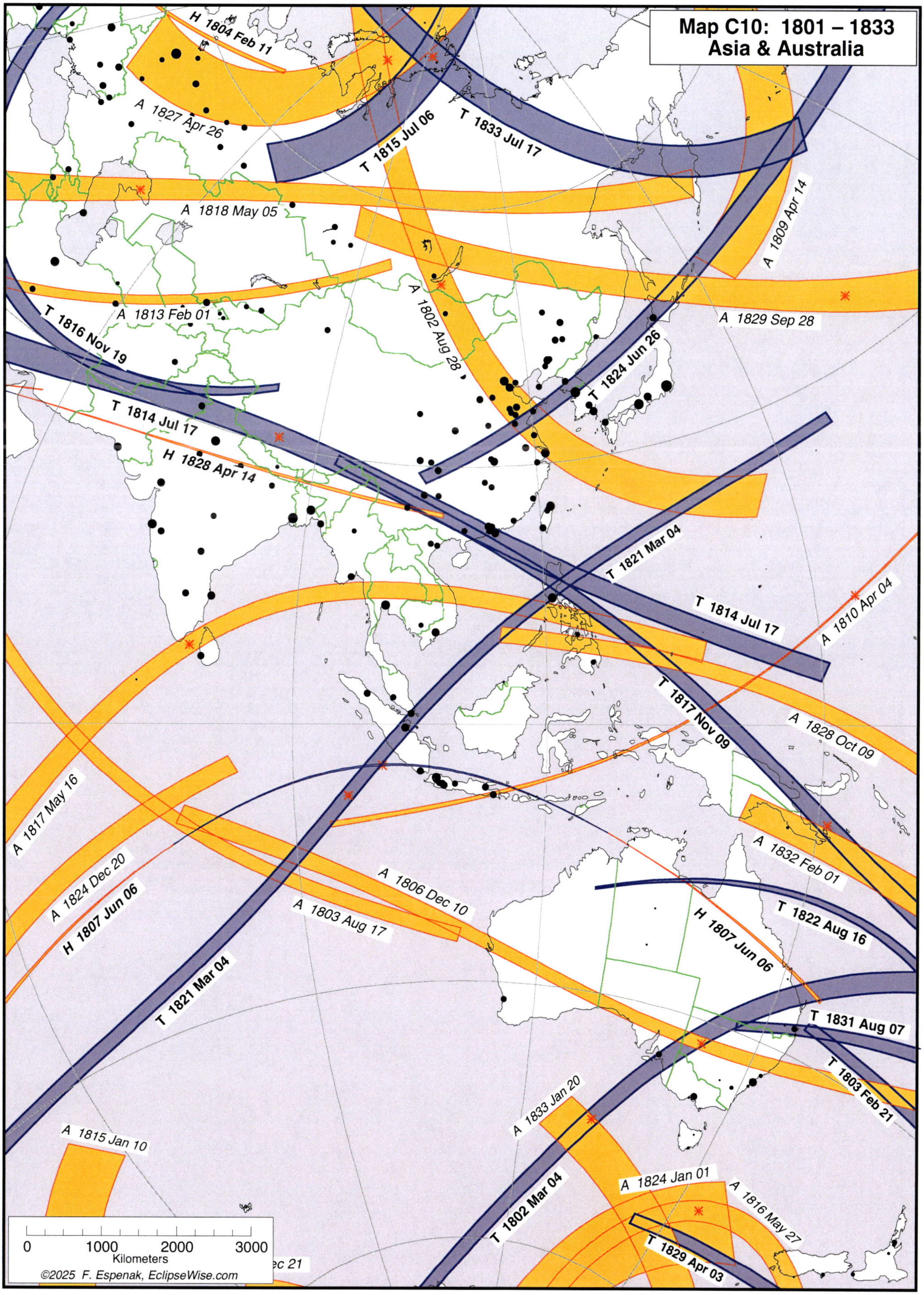
Map C10: 1801 – 1833
Asia & Australia
H 1804 Feb 11
A 1827 Apr 26
T 1815 Jul 06
T 1833 Jul 17
A 1818 May 05
A 1809 Apr 14
A 1813 Feb 01
A 1802 Aug 28
A 1829 Sep 28
T 1816 Nov 19
T 1824 Jun 26
T 1814 Jul 17
H 1828 Apr 14
T 1821 Mar 04
T 1814 Jul 17
A 1810 Apr 04
T 1817 Nov 09
A 1828 Oct 09
A 1817 May 16
A 1832 Feb 01
A 1824 Dec 20
H 1807 Jun 06
A 1806 Dec 10
A 1803 Aug 17
H 1807 Jun 06
T 1822 Aug 16
T 1821 Mar 04
T 1831 Aug 07
T 1803 Feb 21
A 1833 Jan 20
A 1815 Jan 10
A 1824 Jan 01
A 1816 May 27
T 1802 Mar 04
T 1829 Apr 03
0 1000 2000 3000
Kilometers
©2025 F. Espenak, EclipseWise.com

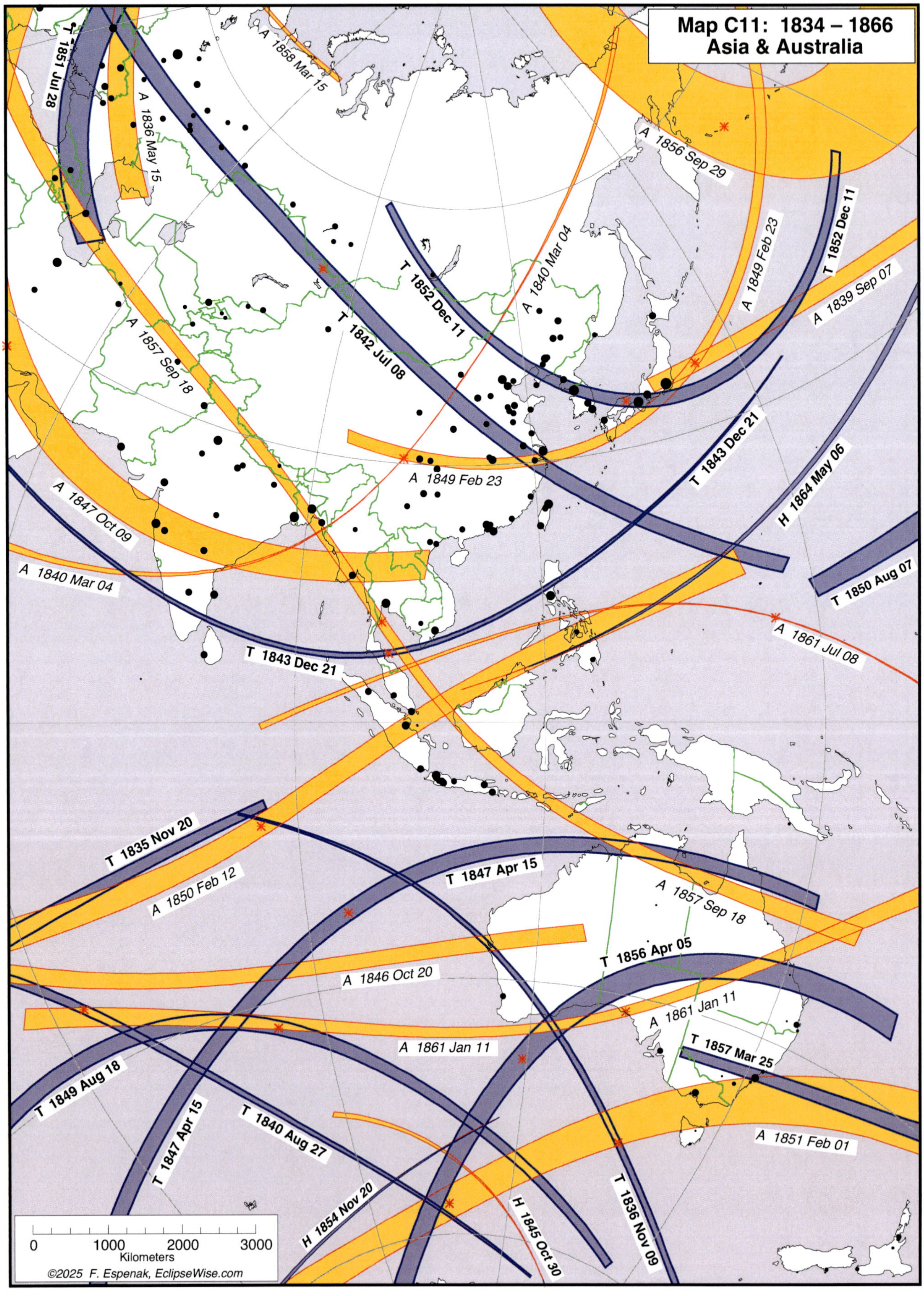
Map C11: 1834 – 1866
Asia & Australia
T 1851 Jul 28
A 1836 May 15
A 1858 Mar 15
A 1856 Sep 29
T 1852 Dec 11
A 1849 Feb 23
A 1840 Mar 04
T 1852 Dec 11
A 1839 Sep 07
A 1857 Sep 18
T 1842 Jul 08
A 1849 Feb 23
T 1843 Dec 21
H 1864 May 06
A 1847 Oct 09
A 1840 Mar 04
T 1850 Aug 07
A 1861 Jul 08
T 1843 Dec 21
T 1835 Nov 20
T 1847 Apr 15
A 1850 Feb 12
A 1857 Sep 18
T 1856 Apr 05
A 1846 Oct 20
A 1861 Jan 11
A 1861 Jan 11
T 1857 Mar 25
T 1849 Aug 18
T 1847 Apr 15
T 1840 Aug 27
A 1851 Feb 01
H 1854 Nov 20
H 1845 Oct 30
T 1836 Nov 09
0 1000 2000 3000
Kilometers
©2025 F. Espenak, EclipseWise.com

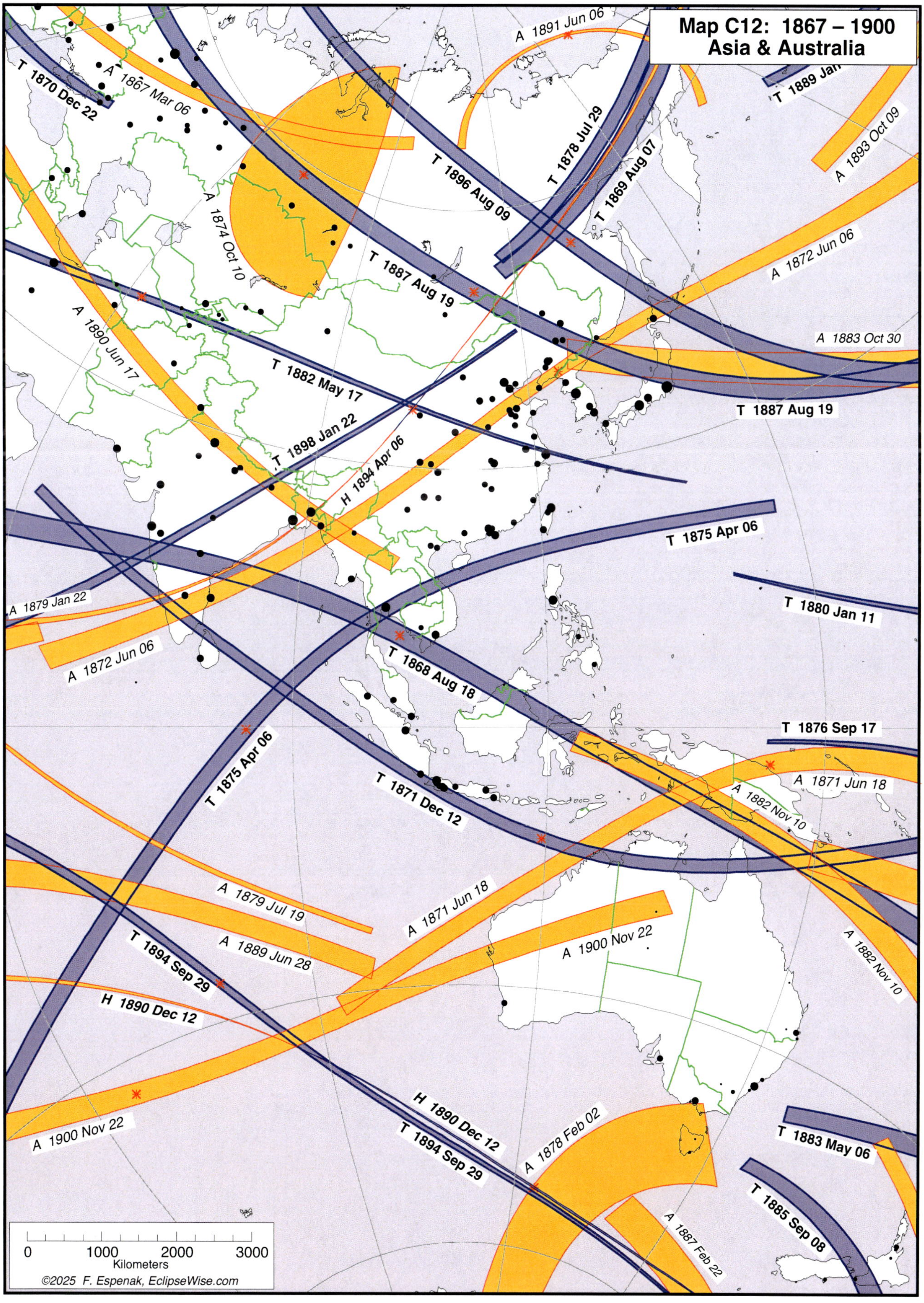
Map C12: 1867 – 1900
Asia & Australia
A 1891 Jun 06
T 1889 Jan
A 1893 Oct 09
T 1870 Dec 22
A 1867 Mar 06
T 1878 Jul 29
T 1869 Aug 07
T 1896 Aug 09
A 1874 Oct 10
T 1887 Aug 19
A 1872 Jun 06
A 1890 Jun 17
A 1883 Oct 30
T 1882 May 17
T 1887 Aug 19
T 1898 Jan 22
H 1894 Apr 06
T 1875 Apr 06
A 1879 Jan 22
T 1880 Jan 11
A 1872 Jun 06
T 1868 Aug 18
T 1876 Sep 17
T 1875 Apr 06
A 1871 Jun 18
T 1871 Dec 12
A 1882 Nov 10
A 1879 Jul 19
A 1871 Jun 18
A 1882 Nov 10
A 1889 Jun 28
A 1900 Nov 22
T 1894 Sep 29
H 1890 Dec 12
A 1900 Nov 22
H 1890 Dec 12
T 1894 Sep 29
A 1878 Feb 02
T 1883 May 06
T 1885 Sep 08
A 1887 Feb 22
0 1000 2000 3000
Kilometers
©2025 F. Espenak, EclipseWise.com

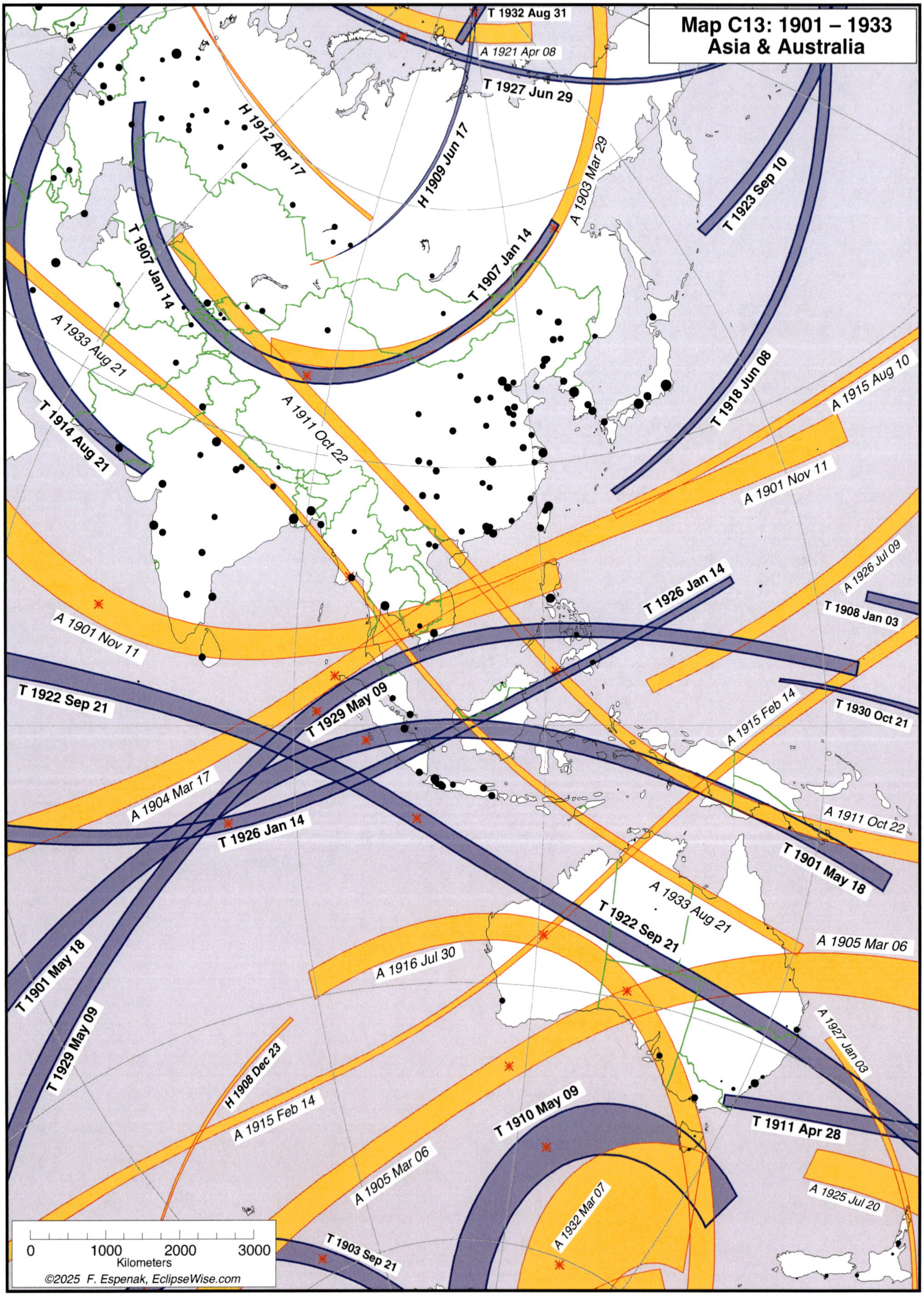
Map C13: 1901 – 1933
Asia & Australia
T 1932 Aug 31
A 1921 Apr 08
T 1927 Jun 29
H 1912 Apr 17
H 1909 Jun 17
A 1903 Mar 29
T 1923 Sep 10
T 1907 Jan 14
T 1907 Jan 14
A 1933 Aug 21
T 1918 Jun 08
A 1915 Aug 10
T 1914 Aug 21
A 1911 Oct 22
A 1901 Nov 11
A 1926 Jul 09
T 1926 Jan 14
T 1908 Jan 03
A 1901 Nov 11
T 1922 Sep 21
T 1929 May 09
T 1930 Oct 21
A 1915 Feb 14
A 1904 Mar 17
T 1926 Jan 14
A 1911 Oct 22
T 1901 May 18
A 1933 Aug 21
T 1922 Sep 21
A 1905 Mar 06
A 1916 Jul 30
T 1901 May 18
A 1927 Jan 03
T 1929 May 09
H 1908 Dec 23
T 1910 May 09
T 1911 Apr 28
A 1915 Feb 14
A 1905 Mar 06
A 1925 Jul 20
A 1932 Mar 07
T 1903 Sep 21
0
1000
2000
3000
Kilometers
©2025 F. Espenak, EclipseWise.com

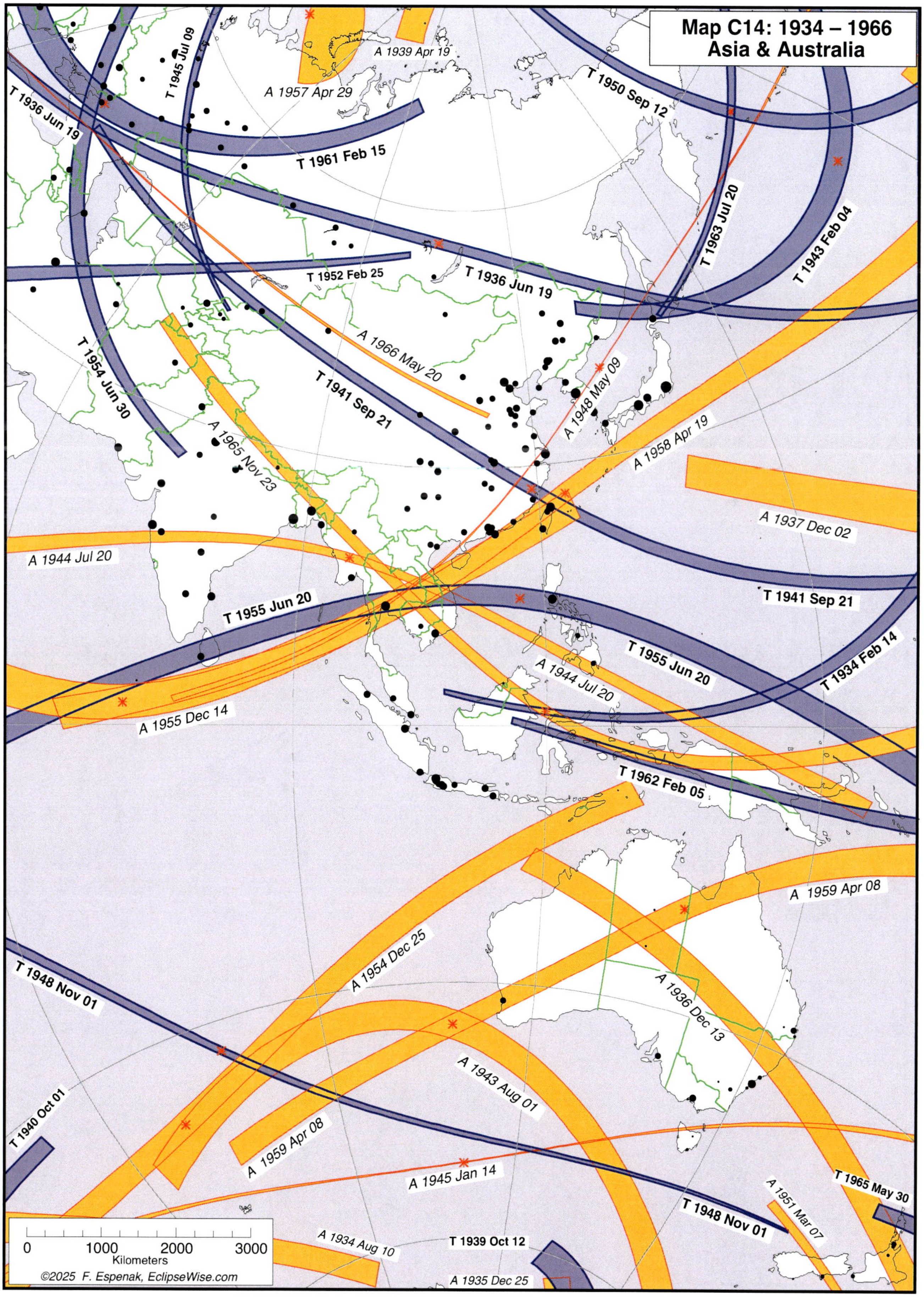
Map C14: 1934 – 1966
Asia & Australia
A 1939 Apr 19
A 1957 Apr 29
T 1945 Jul 09
T 1950 Sep 12
T 1936 Jun 19
T 1961 Feb 15
T 1963 Jul 20
T 1943 Feb 04
T 1952 Feb 25
T 1936 Jun 19
A 1966 May 20
T 1954 Jun 30
T 1941 Sep 21
A 1948 May 09
A 1958 Apr 19
A 1965 Nov 23
A 1937 Dec 02
A 1944 Jul 20
T 1941 Sep 21
T 1955 Jun 20
T 1955 Jun 20
T 1934 Feb 14
A 1944 Jul 20
A 1955 Dec 14
T 1962 Feb 05
A 1959 Apr 08
A 1954 Dec 25
T 1948 Nov 01
A 1936 Dec 13
A 1943 Aug 01
T 1940 Oct 01
A 1959 Apr 08
A 1945 Jan 14
T 1965 May 30
A 1951 Mar 07
T 1948 Nov 01
A 1934 Aug 10
T 1939 Oct 12
A 1935 Dec 25
0 1000 2000 3000
Kilometers
©2025 F. Espenak, EclipseWise.com

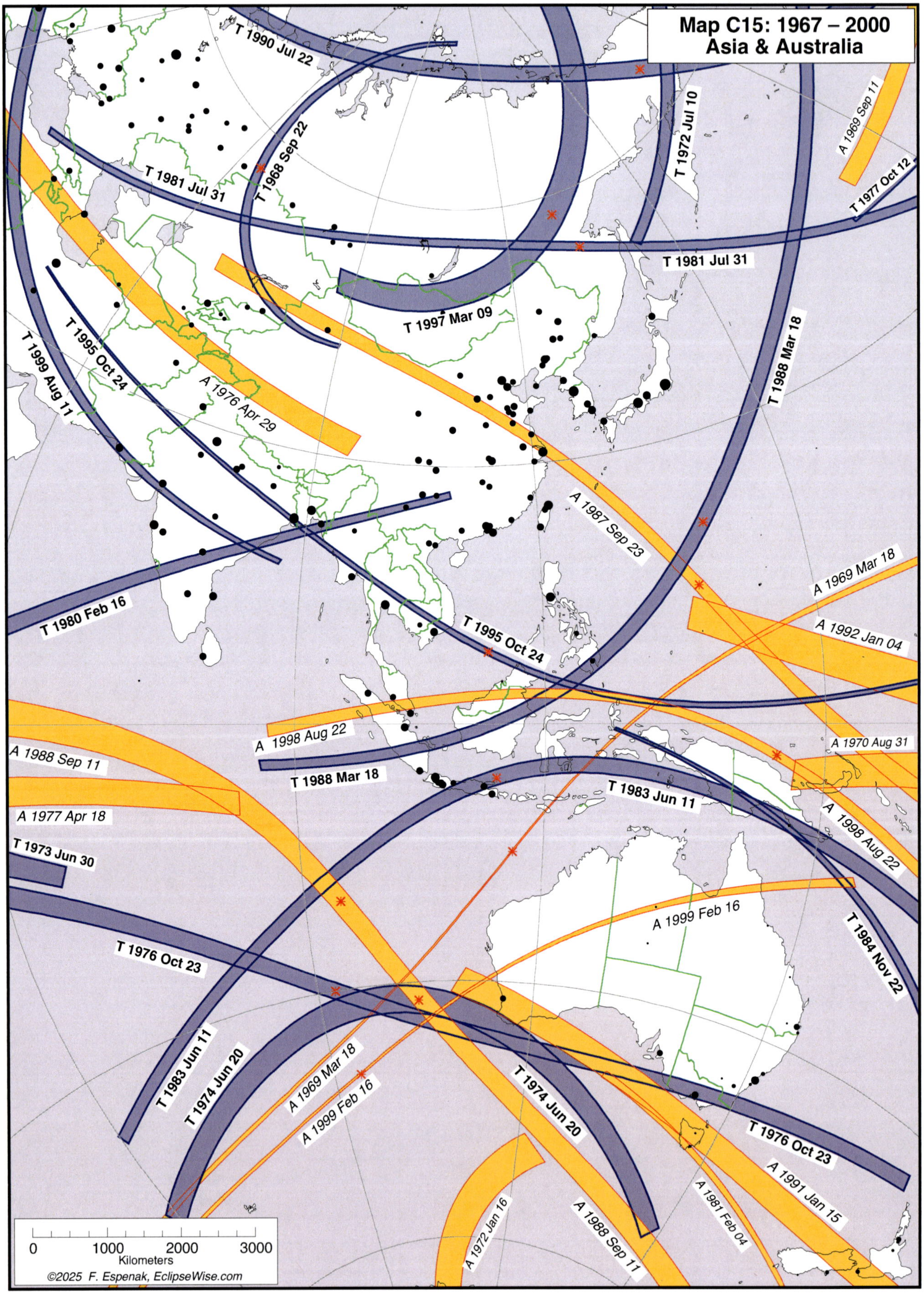
Map C15: 1967 – 2000
Asia & Australia
T 1990 Jul 22
T 1968 Sep 22
T 1972 Jul 10
A 1969 Sep 11
T 1977 Oct 12
T 1981 Jul 31
T 1981 Jul 31
T 1997 Mar 09
T 1988 Mar 18
T 1999 Aug 11
T 1995 Oct 24
A 1976 Apr 29
A 1987 Sep 23
A 1969 Mar 18
A 1992 Jan 04
T 1980 Feb 16
T 1995 Oct 24
A 1998 Aug 22
A 1970 Aug 31
A 1988 Sep 11
T 1988 Mar 18
T 1983 Jun 11
A 1977 Apr 18
A 1998 Aug 22
T 1973 Jun 30
A 1999 Feb 16
T 1984 Nov 22
T 1976 Oct 23
T 1983 Jun 11
T 1974 Jun 20
A 1969 Mar 18
A 1999 Feb 16
T 1974 Jun 20
T 1976 Oct 23
A 1991 Jan 15
A 1981 Feb 04
A 1972 Jan 16
A 1988 Sep 11
0
1000
2000
3000
Kilometers
©2025 F. Espenak, EclipseWise.com

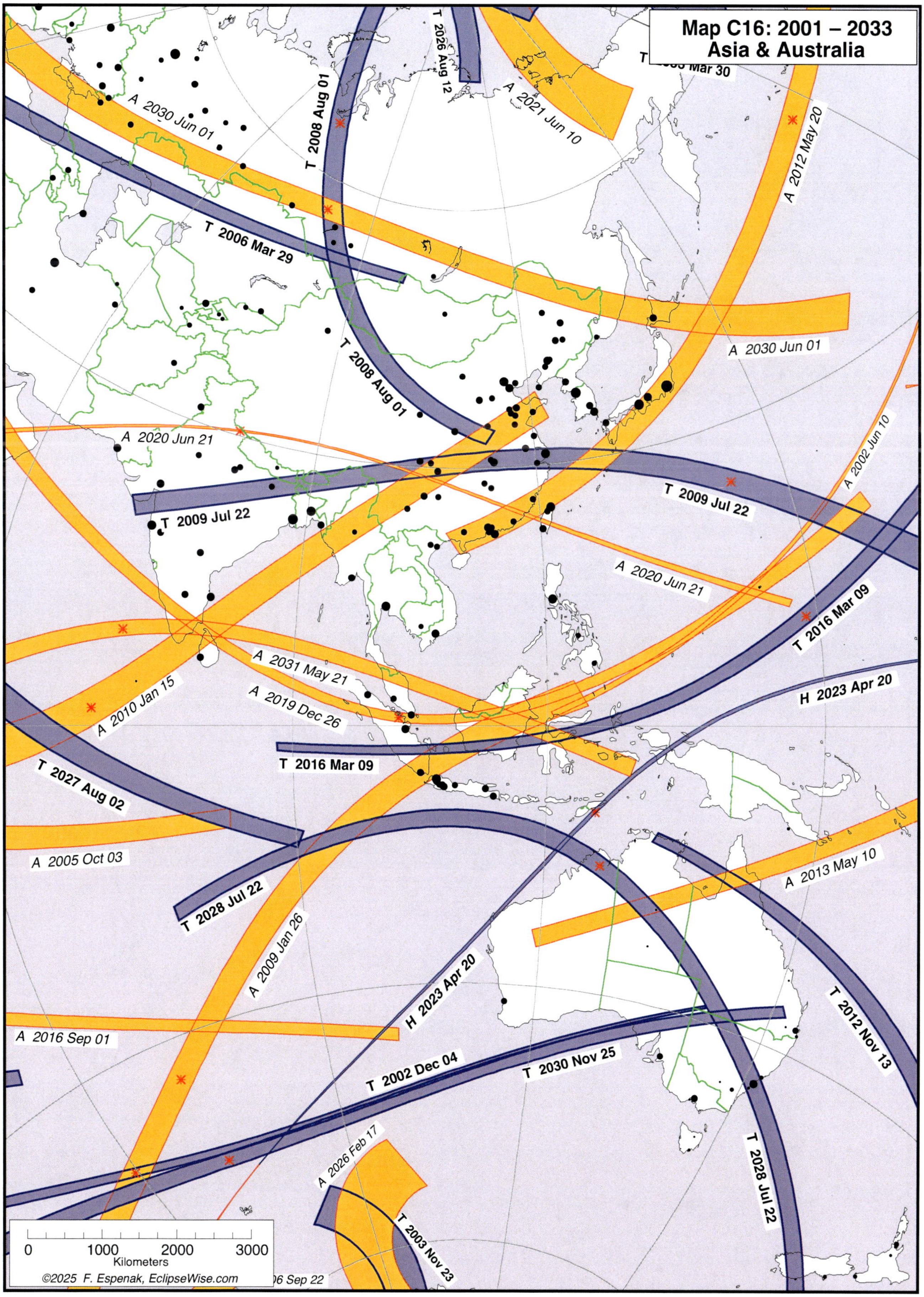
Map C16: 2001 – 2033
Asia & Australia
T 2026 Aug 12
A 2021 Jun 10
A 2030 Jun 01
T 2008 Aug 01
A 2012 May 20
T 2006 Mar 29
A 2030 Jun 01
T 2008 Aug 01
A 2020 Jun 21
A 2002 Jun 10
T 2009 Jul 22
T 2009 Jul 22
A 2020 Jun 21
T 2016 Mar 09
A 2031 May 21
H 2023 Apr 20
A 2019 Dec 26
A 2010 Jan 15
T 2016 Mar 09
T 2027 Aug 02
A 2005 Oct 03
A 2013 May 10
T 2028 Jul 22
A 2009 Jan 26
H 2023 Apr 20
T 2012 Nov 13
A 2016 Sep 01
T 2002 Dec 04
T 2030 Nov 25
A 2026 Feb 17
T 2028 Jul 22
T 2003 Nov 23
0 1000 2000 3000
Kilometers
©2025 F. Espenak, EclipseWise.com

Map C17: 2034 – 2066
Asia & Australia
A 2039 Jun 21
T 2061 Apr 20
T 2043 Apr 09
A 2066 Jun 22
Jun 21
T 2042 Apr 20
A 2057 Jul 01
A 2048 Jun 11
T 2060 Apr 30
T 2034 Mar 20
T 2063 Aug 24
T 2035 Sep 02
A 2041 Oct 25
A 2056 Jan 16
A 2064 Feb 17
T 2042 Apr 20
A 2042 Oct 14
T 2053 Sep 12
H 2049 Nov 25
H 2049 Nov 25
T 2041 Apr 30
A 2059 Nov 05
A 2046 Feb 05
A 2053 Mar 20
T 2038 Dec 26
T 2057 Jan 05
A 2052 Sep 22
A 2042 Oct 14
T 2037 Jul 13
T 2066 Dec 17
A 2063 Feb 28
A 2060 Oct 24
A 2035 Mar 09
A 2045 Feb 16
T 2037 Jul 13
T 2046 Aug 02
T 2055 Jul 24
0
1000
2000
3000
Kilometers
©2025 F. Espenak, EclipseWise.com

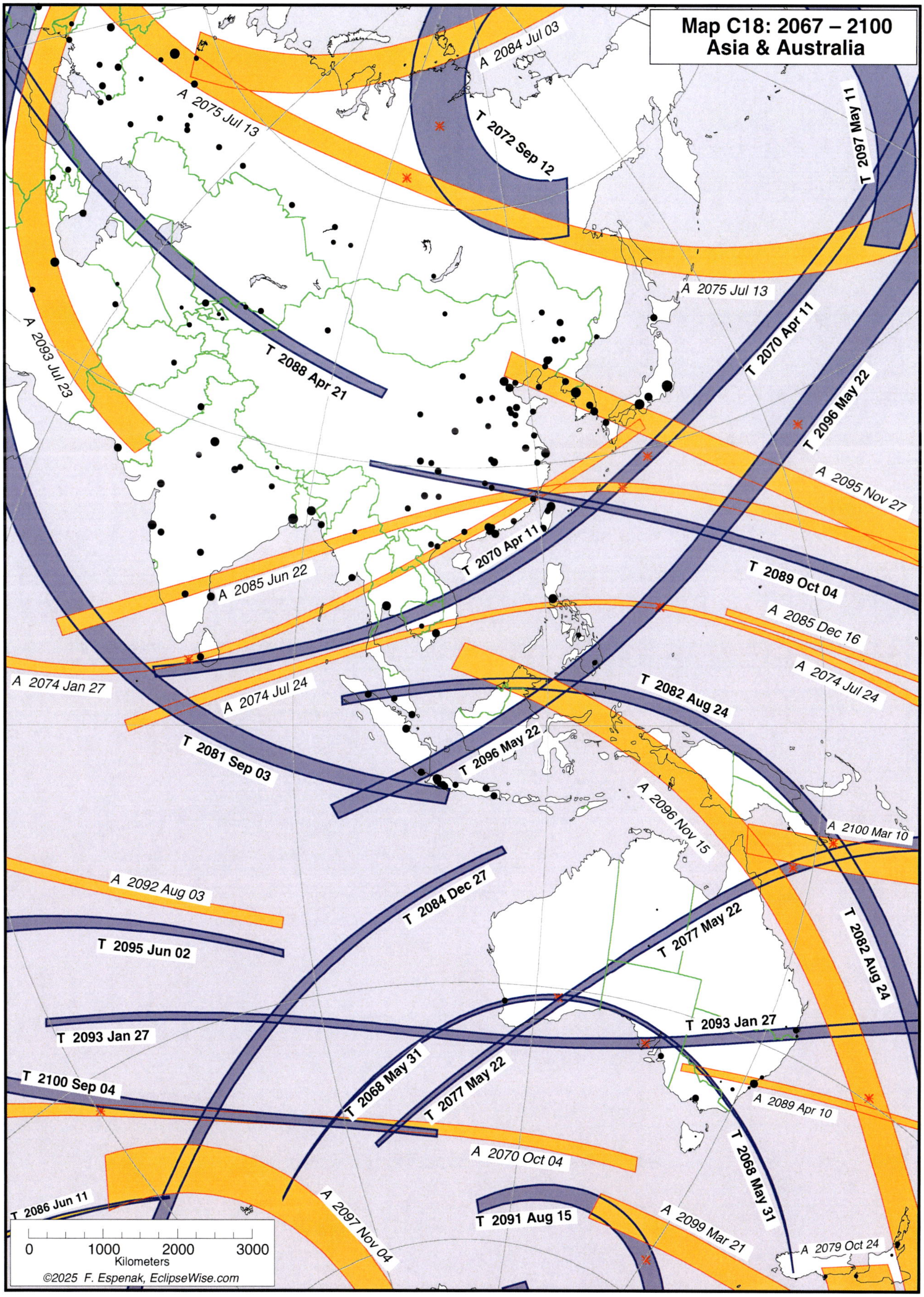
Map C18: 2067 – 2100
Asia & Australia
A 2084 Jul 03
A 2075 Jul 13
T 2072 Sep 12
T 2097 May 11
A 2075 Jul 13
A 2093 Jul 23
T 2088 Apr 21
T 2070 Apr 11
T 2096 May 22
A 2095 Nov 27
T 2070 Apr 11
A 2085 Jun 22
T 2089 Oct 04
A 2085 Dec 16
A 2074 Jan 27
A 2074 Jul 24
A 2074 Jul 24
T 2082 Aug 24
T 2081 Sep 03
T 2096 May 22
A 2096 Nov 15
A 2100 Mar 10
A 2092 Aug 03
T 2084 Dec 27
T 2077 May 22
T 2095 Jun 02
T 2082 Aug 24
T 2093 Jan 27
T 2093 Jan 27
T 2100 Sep 04
T 2068 May 31
T 2077 May 22
A 2089 Apr 10
A 2070 Oct 04
T 2068 May 31
T 2086 Jun 11
A 2097 Nov 04
T 2091 Aug 15
A 2099 Mar 21
A 2079 Oct 24
0 1000 2000 3000
Kilometers
©2025 F. Espenak, EclipseWise.com

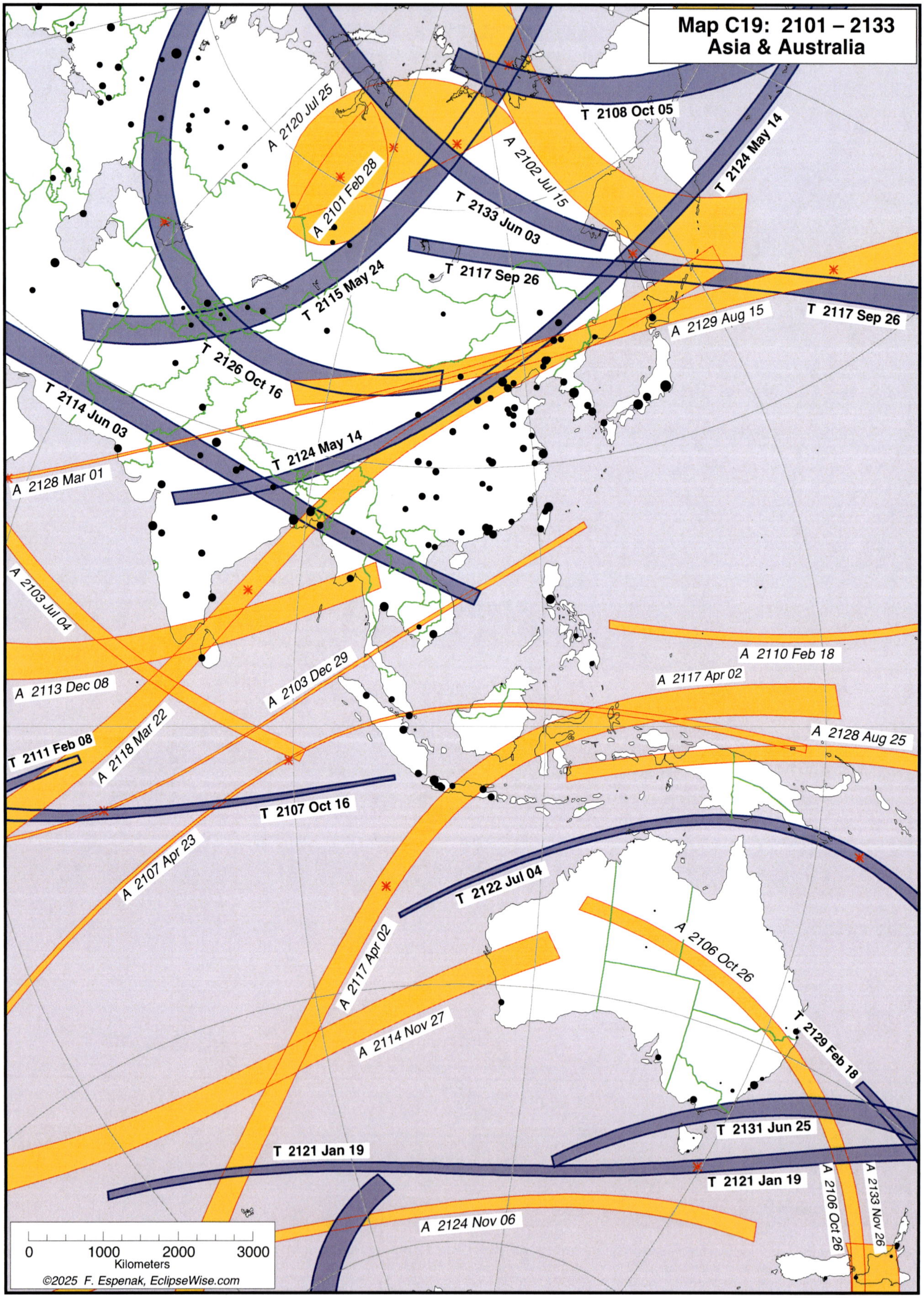
Map C19: 2101 – 2133
Asia & Australia
T 2108 Oct 05
A 2120 Jul 25
A 2102 Jul 15
T 2124 May 14
A 2101 Feb 28
T 2133 Jun 03
T 2117 Sep 26
T 2115 May 24
T 2117 Sep 26
A 2129 Aug 15
T 2126 Oct 16
T 2114 Jun 03
T 2124 May 14
A 2128 Mar 01
A 2103 Jul 04
A 2110 Feb 18
A 2113 Dec 08
A 2103 Dec 29
A 2117 Apr 02
A 2128 Aug 25
T 2111 Feb 08
A 2118 Mar 22
T 2107 Oct 16
A 2107 Apr 23
T 2122 Jul 04
A 2106 Oct 26
A 2117 Apr 02
T 2129 Feb 18
A 2114 Nov 27
T 2131 Jun 25
T 2121 Jan 19
T 2121 Jan 19
A 2106 Oct 26
A 2133 Nov 26
A 2124 Nov 06
0
1000
2000
3000
Kilometers
©2025 F. Espenak, EclipseWise.com

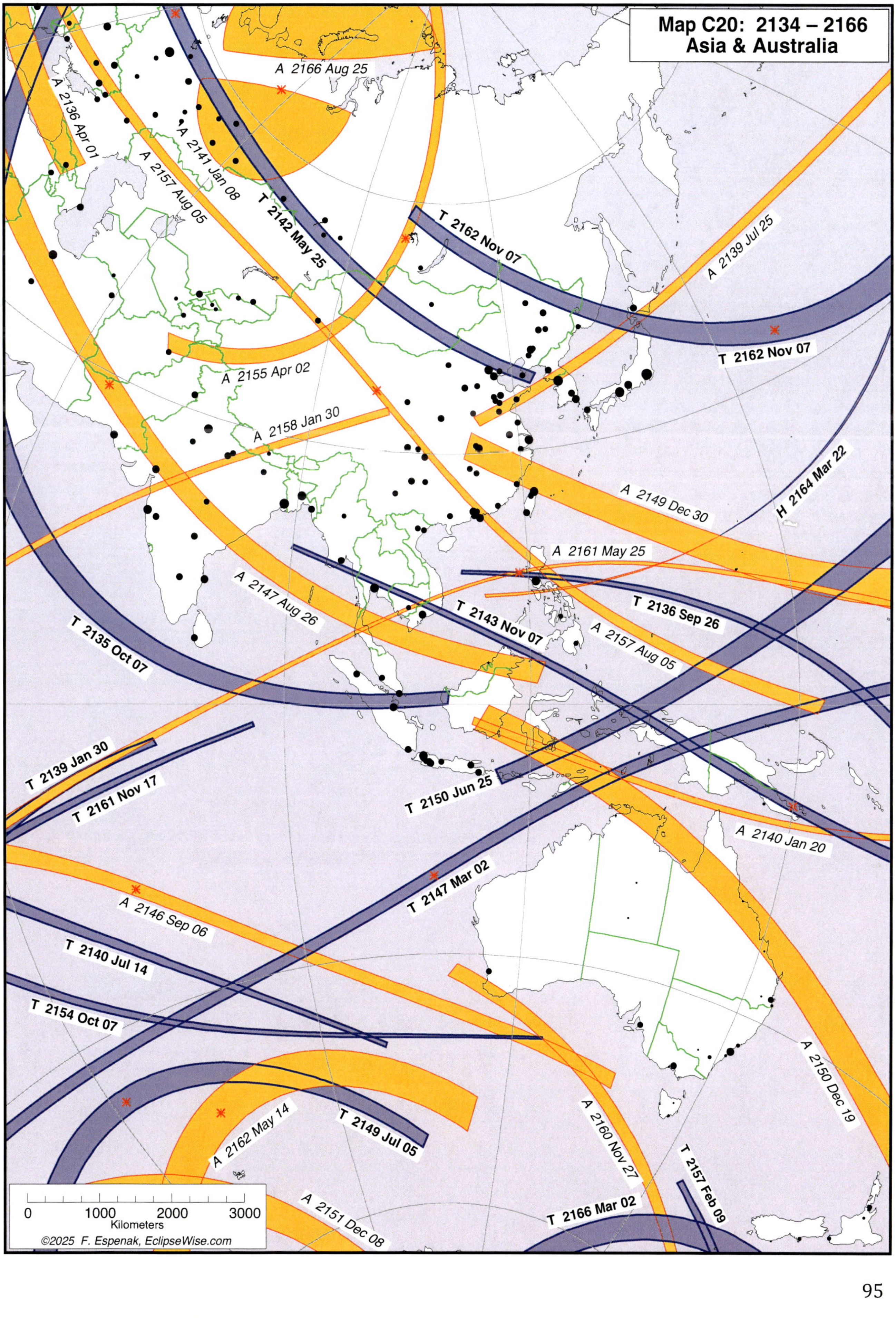
Map C20: 2134 – 2166
Asia & Australia
A 2166 Aug 25
A 2136 Apr 01
A 2141 Jan 08
A 2157 Aug 05
T 2142 May 25
T 2162 Nov 07
A 2139 Jul 25
T 2162 Nov 07
A 2155 Apr 02
A 2158 Jan 30
A 2149 Dec 30
H 2164 Mar 22
A 2161 May 25
A 2147 Aug 26
T 2143 Nov 07
T 2136 Sep 26
A 2157 Aug 05
T 2135 Oct 07
T 2139 Jan 30
T 2161 Nov 17
T 2150 Jun 25
A 2140 Jan 20
T 2147 Mar 02
A 2146 Sep 06
T 2140 Jul 14
T 2154 Oct 07
A 2150 Dec 19
A 2162 May 14
T 2149 Jul 05
A 2160 Nov 27
T 2157 Feb 09
T 2166 Mar 02
A 2151 Dec 08
0
1000
2000
3000
Kilometers
©2025 F. Espenak, EclipseWise.com

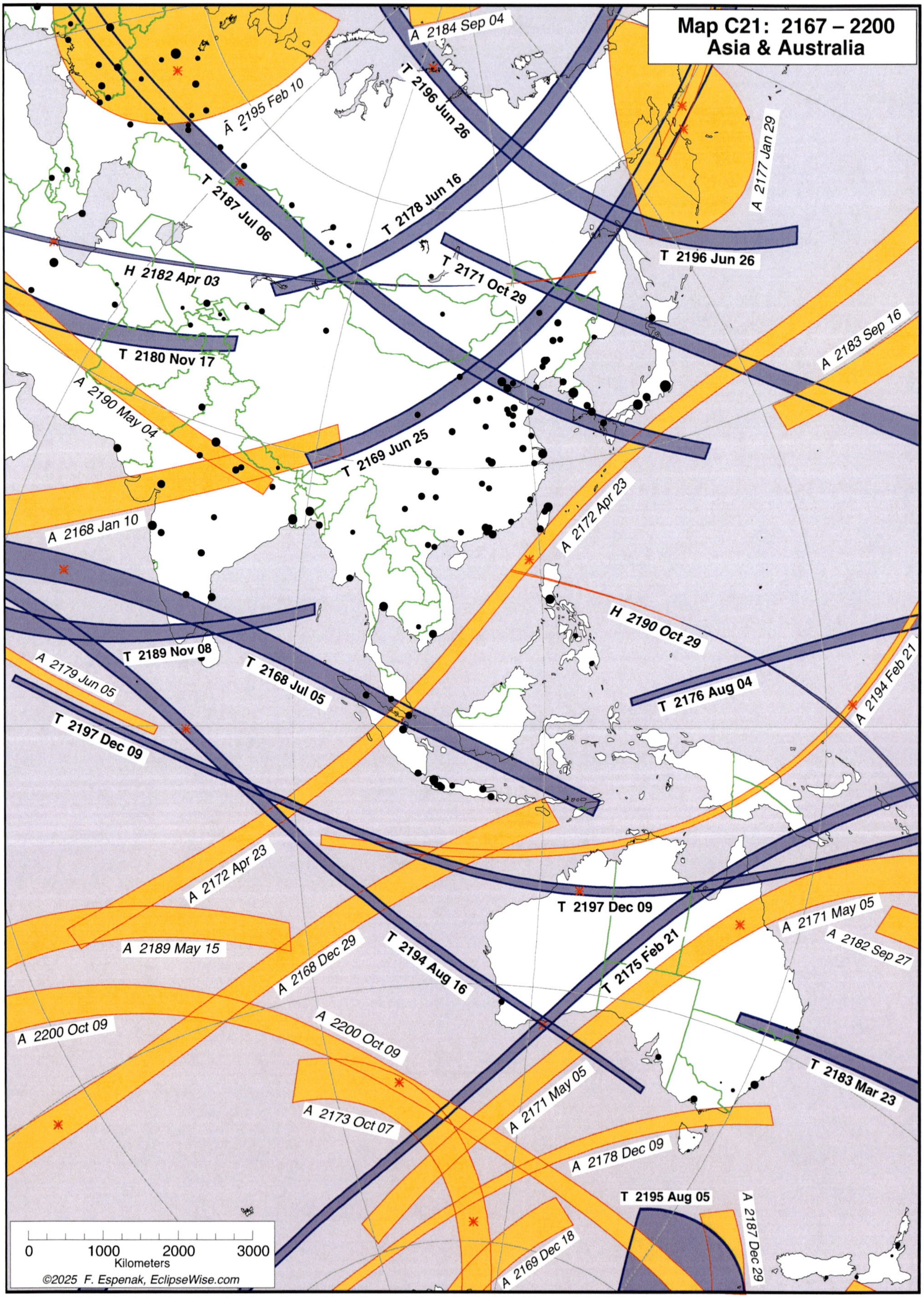
Map C21: 2167 – 2200
Asia & Australia
A 2184 Sep 04
A 2195 Feb 10
T 2196 Jun 26
A 2177 Jan 29
T 2187 Jul 06
T 2178 Jun 16
T 2196 Jun 26
H 2182 Apr 03
T 2171 Oct 29
T 2180 Nov 17
A 2183 Sep 16
A 2190 May 04
T 2169 Jun 25
A 2168 Jan 10
A 2172 Apr 23
H 2190 Oct 29
T 2189 Nov 08
T 2168 Jul 05
A 2179 Jun 05
T 2176 Aug 04
A 2194 Feb 21
T 2197 Dec 09
A 2172 Apr 23
T 2197 Dec 09
A 2171 May 05
A 2182 Sep 27
A 2189 May 15
T 2175 Feb 21
A 2168 Dec 29
T 2194 Aug 16
A 2200 Oct 09
A 2200 Oct 09
T 2183 Mar 23
A 2171 May 05
A 2173 Oct 07
A 2178 Dec 09
T 2195 Aug 05
A 2187 Dec 29
A 2169 Dec 18
0 1000 2000 3000
Kilometers
©2025 F. Espenak, EclipseWise.com

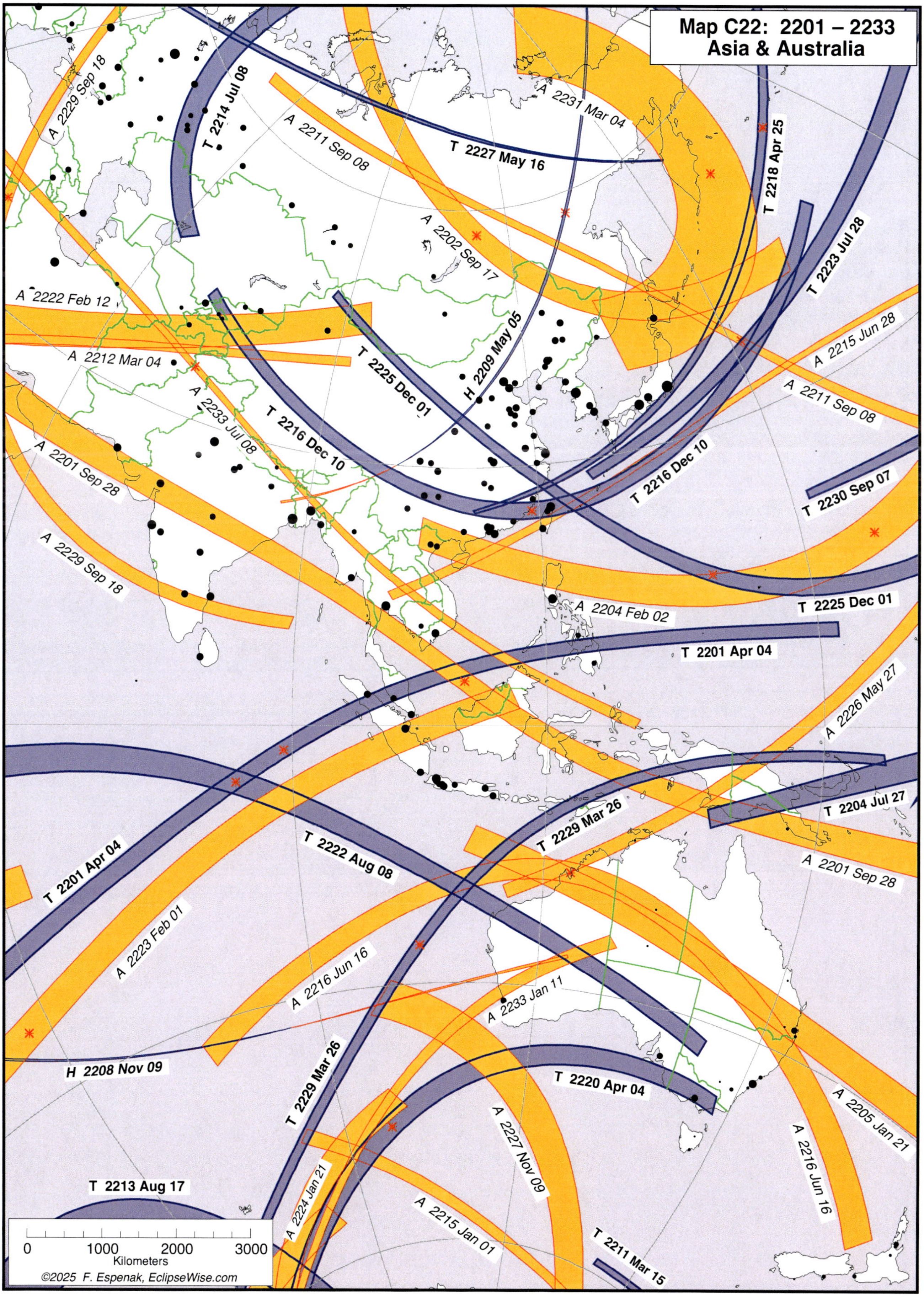
Map C22: 2201 – 2233
Asia & Australia
A 2229 Sep 18
T 2214 Jul 08
A 2211 Sep 08
A 2231 Mar 04
T 2227 May 16
T 2218 Apr 25
A 2202 Sep 17
T 2223 Jul 28
A 2222 Feb 12
H 2209 May 05
T 2225 Dec 01
A 2215 Jun 28
A 2212 Mar 04
A 2233 Jul 08
T 2216 Dec 10
A 2211 Sep 08
A 2201 Sep 28
T 2216 Dec 10
T 2230 Sep 07
A 2229 Sep 18
A 2204 Feb 02
T 2225 Dec 01
T 2201 Apr 04
A 2226 May 27
T 2204 Jul 27
T 2229 Mar 26
T 2201 Apr 04
T 2222 Aug 08
A 2201 Sep 28
A 2223 Feb 01
A 2216 Jun 16
A 2233 Jan 11
H 2208 Nov 09
T 2220 Apr 04
T 2229 Mar 26
A 2227 Nov 09
A 2205 Jan 21
A 2216 Jun 16
T 2213 Aug 17
A 2224 Jan 21
A 2215 Jan 01
T 2211 Mar 15
0
1000
2000
3000
Kilometers
©2025 F. Espenak, EclipseWise.com

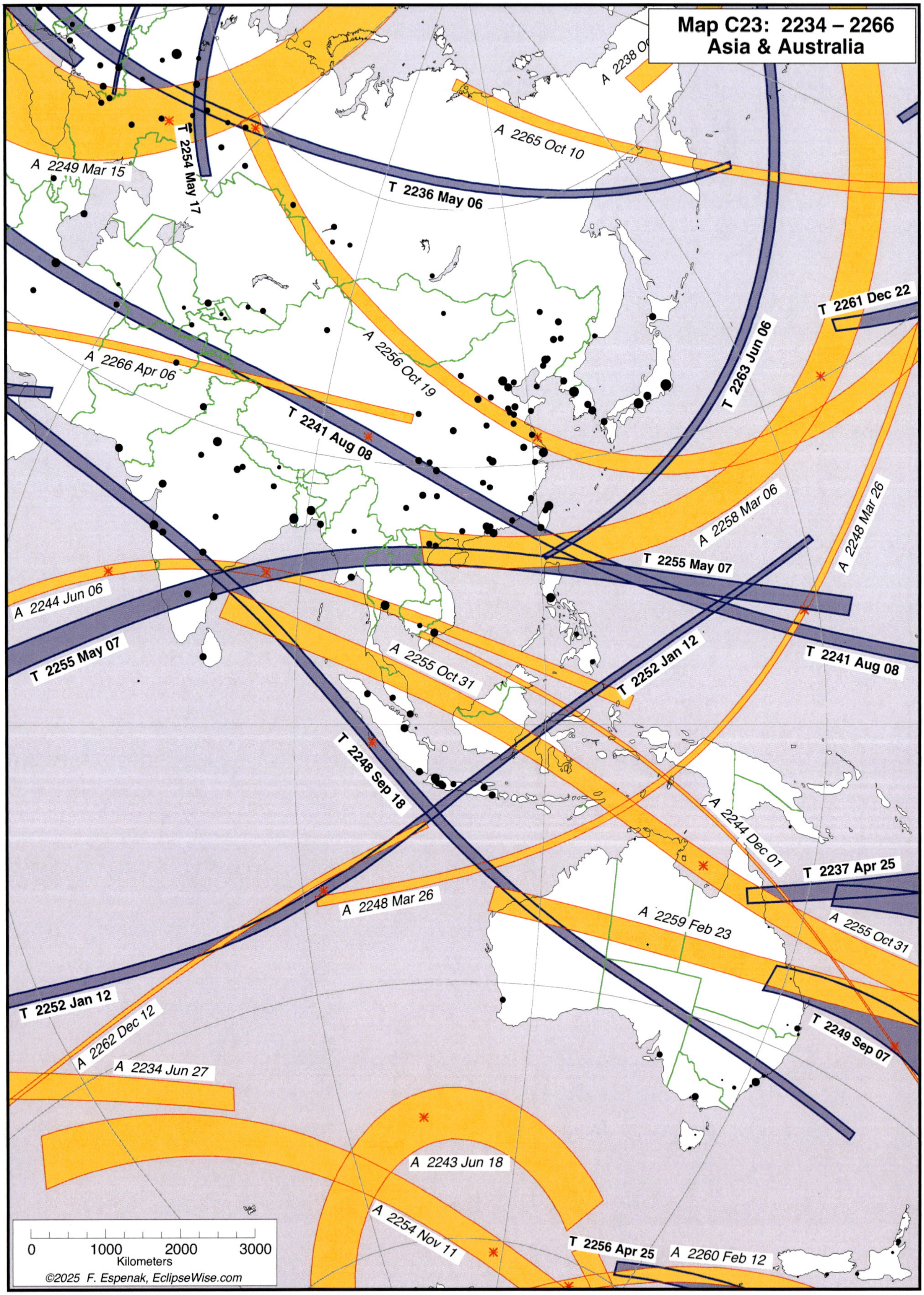
Map C23: 2234 – 2266
Asia & Australia
A 2238 Oct
A 2265 Oct 10
T 2236 May 06
A 2249 Mar 15
T 2254 May 17
A 2256 Oct 19
A 2266 Apr 06
T 2241 Aug 08
T 2261 Dec 22
T 2263 Jun 06
A 2258 Mar 06
A 2248 Mar 26
T 2255 May 07
A 2244 Jun 06
T 2255 May 07
T 2241 Aug 08
A 2255 Oct 31
T 2252 Jan 12
T 2248 Sep 18
A 2244 Dec 01
T 2237 Apr 25
A 2248 Mar 26
A 2259 Feb 23
A 2255 Oct 31
T 2252 Jan 12
T 2249 Sep 07
A 2262 Dec 12
A 2234 Jun 27
A 2243 Jun 18
A 2254 Nov 11
T 2256 Apr 25
A 2260 Feb 12
0 1000 2000 3000
Kilometers
©2025 F. Espenak, EclipseWise.com

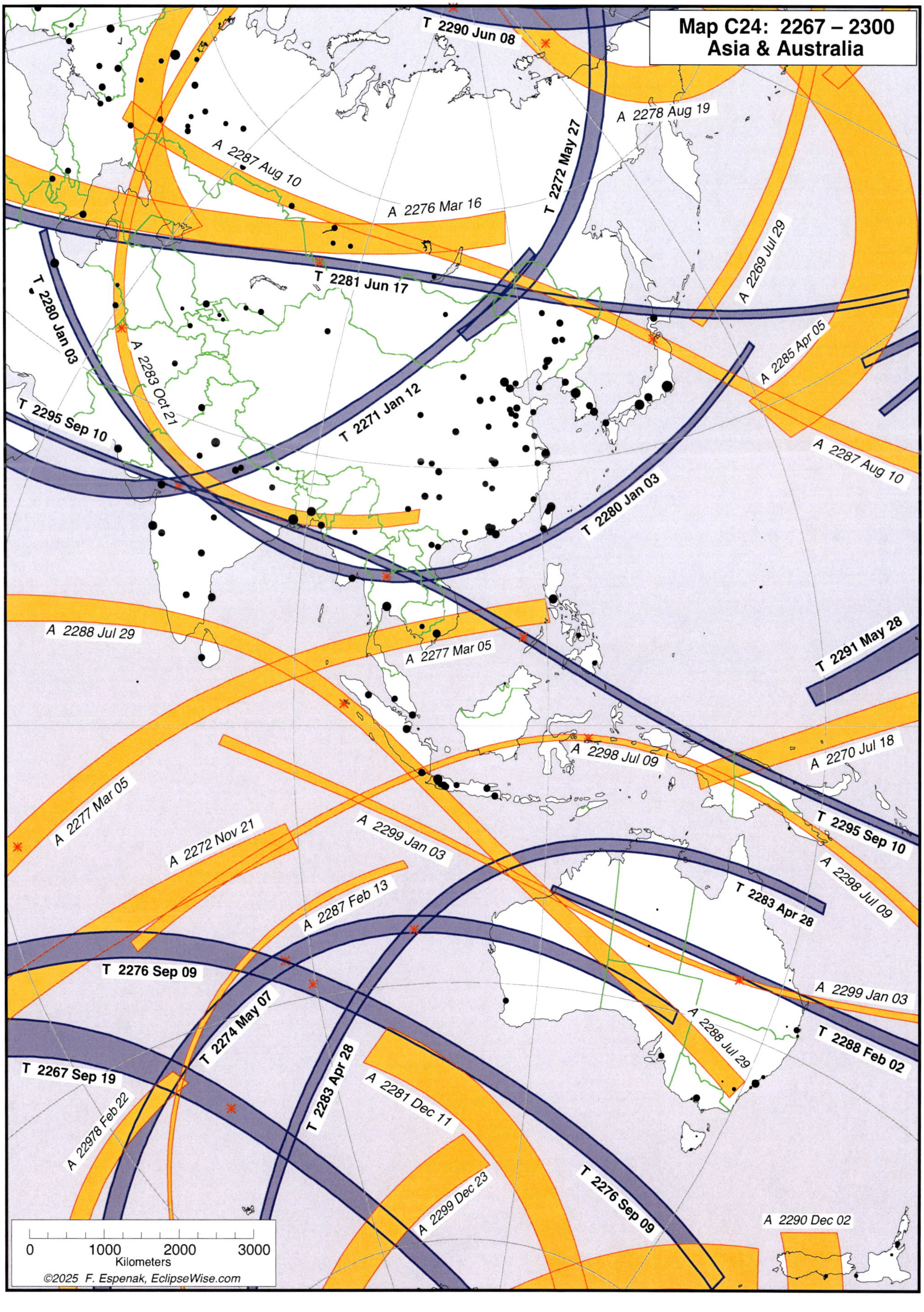
Map C24: 2267 – 2300
Asia & Australia
T 2290 Jun 08
A 2278 Aug 19
T 2272 May 27
A 2287 Aug 10
A 2276 Mar 16
A 2269 Jul 29
T 2281 Jun 17
T 2280 Jan 03
A 2283 Oct 21
A 2285 Apr 05
T 2271 Jan 12
T 2295 Sep 10
A 2287 Aug 10
T 2280 Jan 03
A 2288 Jul 29
A 2277 Mar 05
T 2291 May 28
A 2298 Jul 09
A 2270 Jul 18
A 2277 Mar 05
A 2272 Nov 21
A 2299 Jan 03
T 2295 Sep 10
A 2298 Jul 09
A 2287 Feb 13
T 2283 Apr 28
T 2276 Sep 09
A 2299 Jan 03
T 2274 May 07
A 2288 Jul 29
T 2288 Feb 02
T 2267 Sep 19
T 2283 Apr 28
A 2281 Dec 11
A 22978 Feb 22
T 2276 Sep 09
A 2299 Dec 23
A 2290 Dec 02
0
1000
2000
3000
Kilometers
©2025 F. Espenak, EclipseWise.com

Map C25: 2301 – 2333
Asia & Australia
A 2321 Apr 28
A 2314 Sep 10
A 2330 Apr 19
A 2319 Nov 13
T 2325 Feb 14
T 2308 Jun 19
A 2303 Apr 18
A 2332 Sep 21
A 2303 Apr 18
A 2310 Nov 22
A 2312 Apr 07
T 2309 Jun 09
T 2316 Jan 25
A 2320 May 09
T 2302 Oct 22
T 2306 Feb 14
A 2305 Aug 21
T 2307 Feb 04
T 2309 Jun 09
A 2310 Nov 22
A 2331 Apr 08
A 2317 Jan 14
A 2302 Apr 29
T 2306 Feb 14
T 2328 Jun 09
A 2309 Dec 03
T 2302 Oct 22
T 2303 Oct 11
A 2327 Dec 14
A 2313 Mar 27
T 2328 Jun 09
A 2316 Jul 20
T 2312 Oct 01
T 2321 Oct 22
A 2326 Dec 25
A 2314 Mar 17
T 2310 May 29
T 2330 Oct 13
H 2323 Mar 08
14
0 1000 2000 3000
Kilometers
©2025 F. Espenak, EclipseWise.com

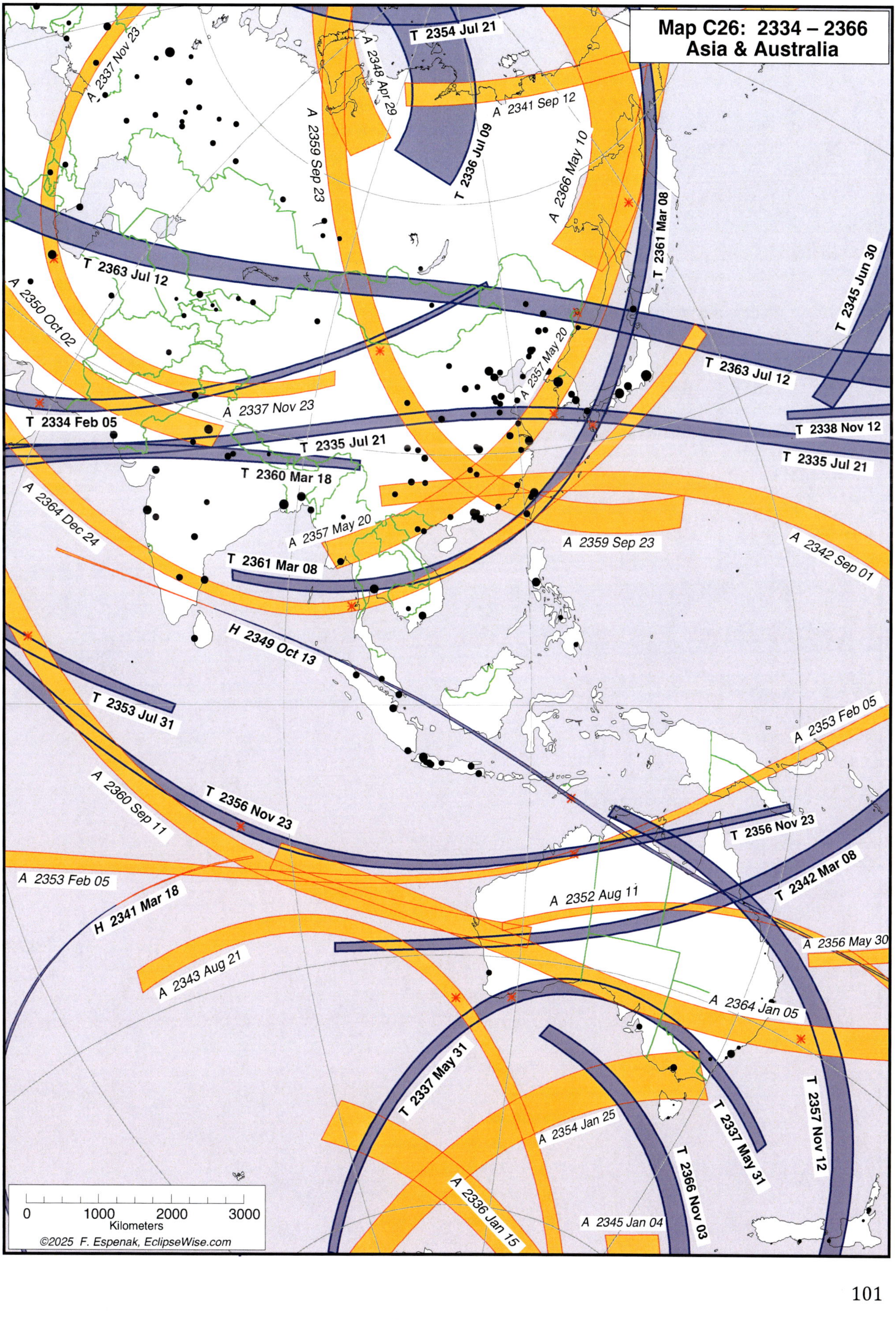
Map C26: 2334 – 2366
Asia & Australia
T 2354 Jul 21
A 2337 Nov 23
A 2348 Apr 29
A 2341 Sep 12
A 2359 Sep 23
T 2336 Jul 09
A 2366 May 10
T 2361 Mar 08
T 2345 Jun 30
T 2363 Jul 12
A 2350 Oct 02
A 2357 May 20
T 2363 Jul 12
A 2337 Nov 23
T 2334 Feb 05
T 2338 Nov 12
T 2335 Jul 21
T 2335 Jul 21
T 2360 Mar 18
A 2364 Dec 24
A 2357 May 20
A 2359 Sep 23
A 2342 Sep 01
T 2361 Mar 08
H 2349 Oct 13
T 2353 Jul 31
A 2353 Feb 05
A 2360 Sep 11
T 2356 Nov 23
T 2356 Nov 23
T 2342 Mar 08
A 2353 Feb 05
A 2352 Aug 11
H 2341 Mar 18
A 2356 May 30
A 2343 Aug 21
A 2364 Jan 05
T 2337 May 31
A 2354 Jan 25
T 2337 May 31
T 2357 Nov 12
T 2366 Nov 03
A 2336 Jan 15
A 2345 Jan 04
0
1000
2000
3000
Kilometers
©2025 F. Espenak, EclipseWise.com

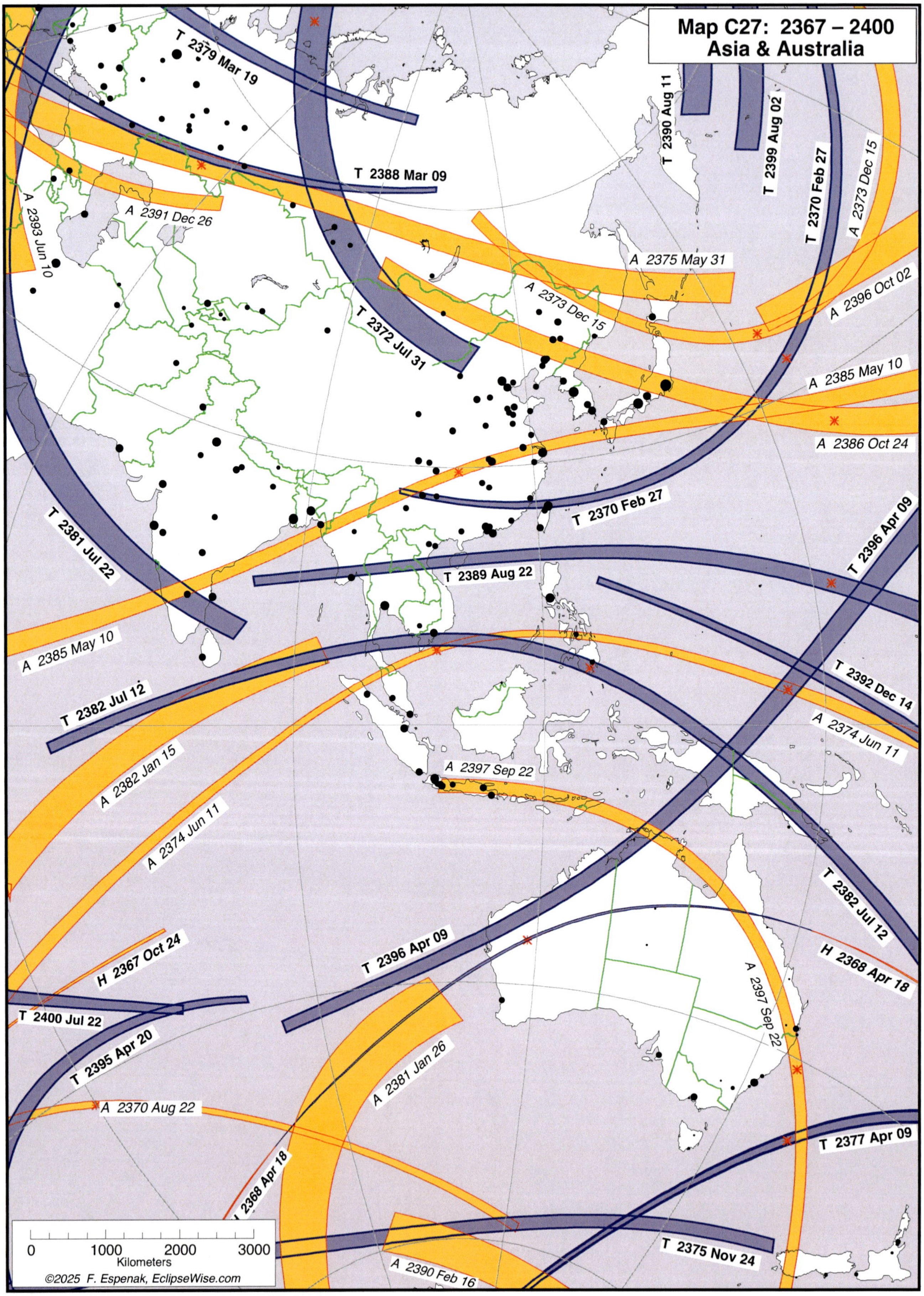
Map C27: 2367 – 2400
Asia & Australia
T 2379 Mar 19
T 2388 Mar 09
T 2390 Aug 11
T 2399 Aug 02
T 2370 Feb 27
A 2373 Dec 15
A 2393 Jun 10
A 2391 Dec 26
A 2375 May 31
A 2373 Dec 15
A 2396 Oct 02
T 2372 Jul 31
A 2385 May 10
A 2386 Oct 24
T 2370 Feb 27
T 2381 Jul 22
T 2389 Aug 22
T 2396 Apr 09
A 2385 May 10
T 2392 Dec 14
T 2382 Jul 12
A 2374 Jun 11
A 2382 Jan 15
A 2397 Sep 22
A 2374 Jun 11
T 2382 Jul 12
T 2396 Apr 09
H 2367 Oct 24
H 2368 Apr 18
A 2397 Sep 22
T 2400 Jul 22
T 2395 Apr 20
A 2381 Jan 26
A 2370 Aug 22
T 2377 Apr 09
2368 Apr 18
T 2375 Nov 24
A 2390 Feb 16
0 1000 2000 3000
Kilometers
©2025 F. Espenak, EclipseWise.com

Appendix D

Catalog of Central Solar Eclipses: 1501 – 2400

Key to Central Solar Eclipse Catalog

Cat Num — sequential Catalog Number assigned to each eclipse from 1 to 1395

Calendar Date — Gregorian date of Greatest Eclipse (Julian date prior to 1582 Oct 04)

TD of Greatest Eclipse — Terrestrial Dynamical Time of Greatest Eclipse

ΔT — arithmetic difference between Terrestrial Dynamical Time Universal Time (seconds)

Saros Num — Saros Series Number of eclipse

Ecl Type — Solar Eclipse Type

A = Annular Solar Eclipse
T = Total Solar Eclipse
H = Hybrid Solar Eclipse

m = Middle eclipse of Saros series
n = Central eclipse of with no northern limit
s = Central eclipse of with no southern limit
+ = Non-central eclipse of with no northern limit
– = Non-central eclipse of with no southern limit
2 = Hybrid eclipse path begins total and ends annular
3 = Hybrid eclipse path begins annular and ends total

Gamma — minimum distance from the axis of the lunar shadow to the center of Earth

Ecl Mag — Eclipse Magnitude; fraction of the Sun's diameter obscured by the Moon

Lat & Long — latitude and longitude where the Sun appears in zenith at greatest eclipse

Sun Alt & **Sun Azm** — altitude and azimuth of the Sun at greatest eclipse

Path Width — width of the central path (km) at greatest eclipse (total, annular & hybrid eclipses)

Central Line Dur — Central Line Duration (minutes, seconds) at greatest eclipse

Cat Num	Calendar Date	TD of Greatest Eclipse	ΔT s	Saros Num	Ecl Type	Gamma	Ecl Mag	Lat °	Long °	Sun Alt °	Sun Azm °	Path Width km	Central Line Dur
0001	1502 Apr 07	05:49:58	288	111	T	-0.4275	1.0567	12.6S	105.1E	65	333	205	04m49s
0002	1502 Oct 01	08:36:15	287	116	A	0.4913	0.9277	19.3N	63.2E	60	210	306	08m16s
0003	1503 Mar 27	22:28:19	286	121	T	0.2905	1.0640	21.1N	163.7W	73	150	218	05m04s
0004	1503 Sep 20	08:52:37	285	126	A	-0.2315	0.9494	14.1S	38.9E	77	30	190	05m27s
0005	1504 Sep 08	15:12:13	283	136	A	-0.9487	0.9924	55.3S	102.2W	18	58	83	00m32s
0006	1505 Jul 30	20:51:54	281	108	T	0.9352	1.0635	67.9N	55.3W	20	263	593	03m25s
0007	1506 Jan 24	05:58:05	280	113	A	-0.3978	0.9209	38.3S	106.5E	66	332	325	08m26s
0008	1506 Jul 20	13:46:56	279	118	T	0.2112	1.0623	30.3N	19.4W	78	202	209	05m08s
0009	1507 Jan 13	07:20:08	278	123	A	0.3024	0.9526	3.0S	68.7E	72	162	181	05m42s
0010	1507 Jul 10	03:06:32	278	128	H	-0.5680	1.0095	12.4S	126.4E	55	16	40	01m01s
0011	1508 Jan 02	15:45:07	277	133	A	0.9733	0.9941	52.8N	76.7W	13	157	92	00m28s
0012	1509 May 18	19:49:34	274	110	A	0.6865	0.9539	64.9N	123.9W	46	171	233	03m56s
0013	1509 Nov 12	10:00:13	273	115	H	-0.4337	1.0131	45.8S	23.6E	64	9	50	01m06s
0014	1510 May 08	01:16:14	272	120	H	-0.1030	1.0033	13.5N	161.8E	84	352	12	00m22s
0015	1510 Nov 01	19:13:49	271	125	A	0.2782	0.9607	1.5S	108.3W	74	190	148	04m54s
0016	1511 Apr 27	13:47:23	270	130	T	-0.8425	1.0463	40.0S	14.3W	32	346	286	03m50s
0017	1511 Oct 21	21:19:48	269	135	A+	1.0059	0.9415	70.7N	98.3W	0	223	-	-
0018	1513 Mar 07	11:54:02	267	112	H	0.3421	1.0040	17.6N	1.5W	70	161	15	00m24s
0019	1513 Aug 30	13:35:50	266	117	T	-0.4392	1.0211	19.5S	31.5W	64	18	80	02m03s
0020	1514 Feb 24	18:45:29	265	122	A	-0.3974	0.9479	28.2S	89.7W	66	341	208	05m51s
0021	1514 Aug 20	04:25:14	264	127	T	0.3032	1.0667	26.5N	120.2E	72	197	228	05m38s
0022	1516 Jun 30	01:15:14	260	109	A	-0.8290	0.9899	33.8S	161.7E	34	2	64	01m03s
0023	1516 Dec 23	14:00:49	259	114	T	0.4256	1.0199	2.2N	27.6W	65	181	75	02m05s
0024	1517 Jun 19	05:41:29	258	119	A	-0.0683	0.9552	19.5N	96.4E	86	357	164	05m50s
0025	1517 Dec 13	05:04:11	258	124	T	-0.2520	1.0468	38.2S	103.3E	75	6	161	03m52s
0026	1518 Jun 08	06:15:22	257	129	A	0.6956	0.9496	67.0N	73.7E	46	162	259	04m13s
0027	1518 Dec 02	20:14:56	256	134	T	-0.9365	1.0124	80.4S	129.0E	20	111	125	00m41s
0028	1520 Apr 17	13:36:45	253	111	T	-0.4825	1.0609	12.6S	11.8W	61	335	226	05m15s
0029	1520 Oct 11	16:03:18	252	116	A	0.5277	0.9244	17.8N	49.1W	58	208	329	08m57s
0030	1521 Apr 07	06:26:04	251	121	T	0.2415	1.0662	22.8N	77.6E	76	151	222	05m15s
0031	1521 Sep 30	16:21:41	250	126	A	-0.1893	0.9489	16.3S	73.0W	79	29	191	05m30s
0032	1522 Mar 27	21:22:58	249	131	T	0.9947	1.0076	62.0N	128.0E	4	84	350	00m26s
0033	1522 Sep 19	22:57:31	248	136	A	-0.9012	0.9946	53.9S	146.4E	25	55	42	00m23s
0034	1523 Aug 11	04:33:15	247	108	Tn	0.9968	1.0559	62.7N	135.7W	2	294	-	02m44s
0035	1524 Feb 04	13:45:34	246	113	A	-0.4176	0.9235	35.4S	9.0W	65	330	315	08m05s
0036	1524 Jul 30	21:17:38	245	118	T	0.2797	1.0577	30.8N	129.9W	74	206	198	04m40s
0037	1525 Jan 23	15:31:19	244	123	A	0.2898	0.9569	1.2S	54.5W	73	159	163	04m58s
0038	1525 Jul 20	10:11:02	243	128	H	-0.4947	1.0054	9.3S	19.7E	60	19	22	00m35s
0039	1526 Jan 13	00:22:29	242	133	A	0.9644	0.9985	51.0N	149.1E	15	151	19	00m07s
0040	1527 May 30	02:22:59	239	110	A	0.7687	0.9556	73.4N	144.9E	39	180	255	03m28s
0041	1527 Nov 23	18:36:36	238	115	H	-0.4421	1.0089	48.6S	102.2W	64	3	34	00m45s
0042	1528 May 18	08:21:04	238	120	H	-0.0290	1.0085	19.9N	54.9E	88	356	29	00m56s
0043	1528 Nov 12	03:22:57	237	125	A	0.2653	0.9562	4.9S	128.7E	75	186	166	05m36s
0044	1529 May 07	21:19:48	236	130	T	-0.7760	1.0526	31.3S	132.8W	39	351	276	04m38s
0045	1529 Nov 01	05:04:10	235	135	An	0.9846	0.9119	61.7N	123.1E	9	201	-	08m09s
0046	1531 Mar 18	19:47:20	232	112	H	0.3817	1.0036	24.3N	121.8W	67	161	13	00m21s
0047	1531 Sep 10	21:21:50	231	117	T	-0.4856	1.0208	26.4S	150.7W	61	20	81	01m56s
0048	1532 Mar 07	02:21:38	230	122	A	-0.3625	0.9488	21.8S	154.6E	69	341	201	05m59s
0049	1532 Aug 30	12:17:44	229	127	T	0.2500	1.0654	19.3N	0.3E	75	198	221	05m40s
0050	1533 Aug 20	05:03:59	227	137	T	0.9693	1.0479	73.7N	178.6E	13	257	679	02m40s
0051	1534 Jul 11	08:08:45	226	109	A	-0.9103	0.9833	44.9S	52.8E	24	8	144	01m35s
0052	1535 Jan 03	22:45:48	225	114	T	0.4285	1.0228	3.8N	159.8W	65	176	86	02m22s
0053	1535 Jun 30	12:08:18	224	119	A	-0.1565	0.9533	13.5N	0.4W	81	1	173	06m19s
0054	1535 Dec 24	13:56:56	223	124	T	-0.2483	1.0469	37.5S	27.1W	75	1	161	03m55s
0055	1536 Jun 18	12:43:20	222	129	A	0.6080	0.9523	61.0N	13.1W	52	174	220	04m17s
0056	1536 Dec 13	04:59:19	221	134	T	-0.9344	1.0098	84.5S	17.0W	20	125	97	00m33s
0057	1538 Apr 28	21:17:29	218	111	T	-0.5431	1.0645	13.7S	127.4W	57	338	249	05m39s
0058	1538 Oct 22	23:38:40	217	116	A	0.5571	0.9214	16.6N	163.8W	56	205	351	09m41s
0059	1539 Apr 18	14:15:06	217	121	T	0.1854	1.0680	23.7N	38.4W	79	154	225	05m28s
0060	1539 Oct 12	00:01:44	216	126	A	-0.1551	0.9484	18.7S	172.5E	81	28	192	05m35s

Cat Num	Calendar Date	TD of Greatest Eclipse	ΔT s	Saros Num	Ecl Type	Gamma	Ecl Mag	Lat °	Long °	Sun Alt °	Sun Azm °	Path Width km	Central Line Dur
0061	1540 Apr 07	05:04:29	215	131	T	0.9462	1.0115	63.1N	35.0E	18	104	123	00m42s
0062	1540 Sep 30	06:54:10	214	136	A	-0.8621	0.9960	54.6S	29.4E	30	54	27	00m17s
0063	1542 Feb 14	21:27:22	211	113	A	-0.4423	0.9265	32.5S	123.5W	64	328	305	07m44s
0064	1542 Aug 11	04:51:04	210	118	T	0.3454	1.0525	30.6N	118.8E	70	209	184	04m12s
0065	1543 Feb 03	23:38:50	209	123	A	0.2736	0.9617	1.0N	176.8W	74	156	143	04m14s
0066	1543 Jul 31	17:16:22	208	128	H	-0.4229	1.0007	7.3S	86.8W	65	22	3	00m05s
0067	1544 Jan 24	08:57:44	207	133	H	0.9534	1.0035	49.7N	16.2E	17	146	40	00m16s
0068	1545 Jun 09	08:57:27	205	110	A	0.8505	0.9567	81.2N	72.2E	31	208	303	03m06s
0069	1545 Dec 04	03:15:40	204	115	H	-0.4479	1.0051	50.1S	132.3E	63	357	20	00m25s
0070	1546 May 29	15:24:38	203	120	H	0.0470	1.0133	25.7N	50.8W	87	180	46	01m24s
0071	1546 Nov 23	11:35:40	202	125	A	0.2562	0.9521	7.3S	5.2E	75	182	181	06m13s
0072	1547 May 19	04:48:57	201	130	T	-0.7060	1.0581	23.5S	110.9E	45	356	270	05m22s
0073	1547 Nov 12	12:54:22	200	135	A	0.9684	0.9106	55.5N	4.5W	14	191	1420	08m59s
0074	1549 Mar 29	03:30:54	197	112	H	0.4285	1.0029	31.4N	120.3E	64	161	11	00m16s
0075	1549 Sep 21	05:16:22	197	117	T	-0.5256	1.0205	33.2S	88.2E	58	21	82	01m49s
0076	1550 Mar 18	09:47:47	196	122	A	-0.3200	0.9497	15.1S	41.0E	71	342	194	06m05s
0077	1550 Sep 10	20:17:36	195	127	T	0.2029	1.0636	12.4N	121.6W	78	198	212	05m37s
0078	1551 Aug 31	12:53:00	193	137	T	0.9185	1.0460	65.8N	28.6E	23	226	391	02m52s
0079	1552 Jul 21	15:03:46	192	109	A	-0.9892	0.9742	62.9S	64.4W	7	20	724	02m05s
0080	1553 Jan 14	07:28:08	191	114	T	0.4339	1.0263	6.3N	68.5E	64	172	99	02m41s
0081	1553 Jul 10	18:36:33	190	119	A	-0.2430	0.9509	6.8N	98.4W	76	5	185	06m46s
0082	1554 Jan 03	22:49:37	189	124	T	-0.2447	1.0474	35.8S	157.9W	76	355	163	04m00s
0083	1554 Jun 29	19:10:38	188	129	A	0.5192	0.9546	54.0N	104.6W	58	182	195	04m22s
0084	1554 Dec 24	13:45:20	187	134	T	-0.9341	1.0075	87.5S	159.3E	20	176	75	00m25s
0085	1556 May 09	04:53:34	185	111	T	-0.6079	1.0673	16.0S	117.9E	52	342	274	05m58s
0086	1556 Nov 02	07:22:12	184	116	A	0.5798	0.9190	15.5N	79.1E	54	201	370	10m24s
0087	1557 Apr 28	21:59:03	183	121	T	0.1252	1.0692	24.0N	153.0W	83	157	227	05m42s
0088	1557 Oct 22	07:49:26	182	126	A	-0.1267	0.9482	21.1S	56.2E	83	25	192	05m40s
0089	1558 Apr 18	12:39:26	182	131	T	0.8930	1.0132	64.1N	67.7W	26	114	100	00m50s
0090	1558 Oct 11	14:58:54	181	136	A	-0.8290	0.9971	56.5S	90.1W	34	53	18	00m12s
0091	1560 Feb 26	05:00:43	178	113	A	-0.4741	0.9299	29.9S	123.7E	62	327	294	07m22s
0092	1560 Aug 21	12:30:54	177	118	T	0.4050	1.0469	29.7N	5.4E	66	212	170	03m44s
0093	1561 Feb 14	07:39:20	177	123	A	0.2507	0.9670	3.4N	62.8E	75	153	122	03m30s
0094	1561 Aug 11	00:27:06	176	128	A	-0.3564	0.9956	6.5S	165.6E	69	25	16	00m27s
0095	1562 Feb 03	17:27:32	175	133	T	0.9373	1.0091	48.6N	114.3W	20	142	89	00m41s
0096	1563 Jun 20	15:30:53	173	110	A	0.9338	0.9565	81.3N	55.4E	20	290	454	02m49s
0097	1563 Dec 15	11:55:48	172	115	H	-0.4524	1.0020	50.3S	6.9E	63	350	8	00m10s
0098	1564 Jun 08	22:26:47	171	120	H2	0.1252	1.0174	30.8N	155.3W	83	185	60	01m44s
0099	1564 Dec 03	19:52:05	170	125	A	0.2504	0.9487	8.8S	119.0W	76	178	195	06m42s
0100	1565 May 29	12:14:58	169	130	T	-0.6329	1.0629	16.5S	3.6W	51	0	266	05m57s
0101	1565 Nov 22	20:49:54	168	135	A	0.9564	0.9092	51.4N	130.3W	16	184	1221	09m37s
0102	1567 Apr 09	11:04:07	166	112	H	0.4830	1.0020	38.9N	5.1E	61	161	8	00m11s
0103	1567 Oct 02	13:20:26	165	117	T	-0.5584	1.0200	39.7S	34.8W	56	22	82	01m42s
0104	1568 Mar 28	17:04:20	164	122	A	-0.2701	0.9507	8.1S	70.4W	74	343	187	06m10s
0105	1568 Sep 21	04:25:00	163	127	T	0.1619	1.0615	5.8N	114.7E	81	198	204	05m32s
0106	1569 Mar 17	17:21:17	163	132	A-	-1.0034	0.9488	72.1S	3.3E	0	279	-	-
0107	1569 Sep 10	20:48:15	162	137	T	0.8733	1.0428	57.4N	103.2W	29	215	293	02m55s
0108	1571 Jan 25	16:07:35	159	114	T	0.4421	1.0302	9.5N	62.7W	64	169	113	02m59s
0109	1571 Jul 22	01:07:17	158	119	A	-0.3266	0.9481	0.5S	162.2E	71	9	201	07m08s
0110	1572 Jan 15	07:38:11	158	124	T	-0.2381	1.0485	33.0S	71.7E	76	351	166	04m07s
0111	1572 Jul 10	01:42:41	157	129	A	0.4339	0.9562	46.6N	159.8E	64	187	177	04m30s
0112	1573 Jan 03	22:28:34	156	134	H	-0.9328	1.0058	85.9S	53.9W	21	258	57	00m20s
0113	1574 May 20	12:25:41	154	111	T	-0.6762	1.0694	19.7S	3.8E	47	346	305	06m09s
0114	1574 Nov 13	15:12:16	153	116	A	0.5970	0.9171	14.8N	39.9W	53	197	387	11m03s
0115	1575 May 10	05:34:44	152	121	Tm	0.0584	1.0697	23.1N	94.7E	87	162	227	05m56s
0116	1575 Nov 02	15:47:26	151	126	A	-0.1061	0.9483	23.5S	62.5W	84	22	191	05m44s
0117	1576 Apr 28	20:04:43	150	131	T	0.8329	1.0140	64.8N	167.9W	33	124	86	00m55s
0118	1576 Oct 21	23:13:04	149	136	A	-0.8032	0.9981	59.2S	148.0E	36	51	11	00m08s
0119	1578 Mar 08	12:26:50	147	113	A	-0.5120	0.9336	27.7S	12.4E	59	327	283	07m01s
0120	1578 Sep 01	20:15:07	146	118	T	0.4602	1.0408	28.4N	109.5W	62	213	152	03m17s

Cat Num	Calendar Date	TD of Greatest Eclipse	ΔT s	Saros Num	Ecl Type	Gamma	Ecl Mag	Lat °	Long °	Sun Alt °	Sun Azm °	Path Width km	Central Line Dur
0121	1579 Feb 25	15:34:46	145	123	A	0.2230	0.9728	6.0N	56.3W	77	151	100	02m48s
0122	1579 Aug 22	07:41:31	144	128	A	-0.2938	0.9901	6.6S	57.3E	73	27	36	01m00s
0123	1580 Feb 15	01:52:11	144	133	T	0.9165	1.0151	47.9N	117.4E	23	138	127	01m07s
0124	1581 Dec 25	20:35:19	140	115	A	-0.4566	0.9993	49.4S	118.5W	63	343	3	00m04s
0125	1582 Jun 20	05:30:26	140	120	T	0.2031	1.0210	35.0N	100.8E	78	190	73	01m59s
0126	1582 Dec 25	04:08:38	139	125	A	0.2457	0.9459	9.4S	116.9E	76	173	206	07m02s
0127	1583 Jun 19	19:39:31	138	130	T	-0.5581	1.0667	10.4S	116.9W	56	4	262	06m23s
0128	1583 Dec 14	04:48:37	137	135	A	0.9471	0.9083	48.5N	104.2E	18	177	1116	10m03s
0129	1585 Apr 29	18:28:57	135	112	H	0.5435	1.0005	46.6N	107.6W	57	162	2	00m03s
0130	1585 Oct 22	21:33:24	134	117	T	-0.5845	1.0196	45.7S	159.2W	54	21	82	01m35s
0131	1586 Apr 19	00:10:07	133	122	A	-0.2120	0.9517	0.9S	179.0W	78	345	181	06m12s
0132	1586 Oct 12	12:40:31	132	127	T	0.1278	1.0591	0.3S	10.8W	83	197	196	05m23s
0133	1587 Apr 08	00:27:04	131	132	A	-0.9502	0.9271	60.5S	151.9W	18	325	889	06m26s
0134	1587 Oct 02	04:51:23	130	137	T	0.8352	1.0387	50.0N	128.3E	33	208	235	02m51s
0135	1589 Feb 15	00:42:19	128	114	T	0.4544	1.0344	13.6N	167.1E	63	166	129	03m17s
0136	1589 Aug 11	07:41:03	127	119	A	-0.4072	0.9450	8.2S	61.4E	66	12	221	07m24s
0137	1590 Feb 04	16:24:04	126	124	T	-0.2293	1.0498	29.3S	58.8W	77	347	170	04m17s
0138	1590 Jul 31	08:17:38	125	129	A	0.3503	0.9574	38.8N	61.6E	69	191	166	04m38s
0139	1591 Jan 25	07:09:21	125	134	H	-0.9298	1.0047	81.9S	150.6E	21	283	45	00m16s
0140	1592 Jun 09	19:55:48	122	111	T	-0.7465	1.0705	24.7S	110.3W	42	350	344	06m11s
0141	1592 Dec 03	23:07:15	121	116	A	0.6102	0.9159	14.5N	160.2W	52	193	401	11m36s
0142	1593 May 30	13:07:30	120	121	T	-0.0106	1.0696	21.4N	17.0W	90	342	227	06m08s
0143	1593 Nov 22	23:52:05	120	126	A	-0.0906	0.9488	25.4S	177.5E	85	18	189	05m46s
0144	1594 May 20	03:23:16	119	131	T	0.7678	1.0141	64.9N	94.1E	40	136	76	00m58s
0145	1594 Nov 12	07:34:48	118	136	A	-0.7829	0.9991	62.4S	25.1E	38	48	5	00m04s
0146	1596 Mar 28	19:43:17	116	113	A	-0.5582	0.9373	26.3S	96.5W	56	328	275	06m41s
0147	1596 Sep 22	04:07:01	115	118	T	0.5084	1.0346	26.8N	133.0E	59	213	134	02m50s
0148	1597 Mar 17	23:22:38	114	123	A	0.1879	0.9788	8.4N	173.4W	79	151	77	02m08s
0149	1597 Sep 11	15:01:21	113	128	A	-0.2363	0.9843	7.6S	52.4W	76	29	57	01m35s
0150	1598 Mar 07	10:10:00	112	133	T	0.8894	1.0214	47.7N	8.2W	27	135	156	01m33s
0151	1598 Aug 31	18:48:47	111	138	A-	-1.0126	0.9397	61.2S	175.7W	0	72	-	-
0152	1600 Jan 16	05:12:45	109	115	A	-0.4622	0.9972	47.4S	115.9E	62	337	11	00m14s
0153	1600 Jul 10	12:35:57	108	120	T	0.2803	1.0238	38.2N	2.7W	74	196	84	02m08s
0154	1601 Jan 04	12:24:37	107	125	A	0.2410	0.9437	9.1S	7.1W	76	169	214	07m13s
0155	1601 Jun 30	03:03:58	107	130	T	-0.4826	1.0697	5.3S	130.7E	61	8	259	06m37s
0156	1601 Dec 24	12:50:30	106	135	A	0.9402	0.9078	46.6N	21.6W	19	171	1051	10m14s
0157	1603 May 11	01:44:58	104	112	A	0.6107	0.9987	54.7N	142.6E	52	163	6	00m07s
0158	1603 Nov 03	05:54:54	103	117	T	-0.6041	1.0193	51.1S	75.5E	53	20	83	01m31s
0159	1604 Apr 29	07:07:20	103	122	A	-0.1474	0.9525	6.3N	74.7E	82	347	176	06m12s
0160	1604 Oct 22	21:03:47	102	127	T	0.1000	1.0567	5.9S	137.9W	84	195	188	05m12s
0161	1605 Apr 18	07:26:43	101	132	A	-0.8919	0.9327	49.8S	89.8E	26	337	553	06m43s
0162	1605 Oct 12	12:59:58	100	137	T	0.8023	1.0344	43.4N	0.6E	36	203	193	02m43s
0163	1607 Feb 26	09:10:37	98	114	T	0.4727	1.0388	18.4N	38.1E	62	163	147	03m34s
0164	1607 Aug 22	14:20:47	98	119	A	-0.4824	0.9416	16.1S	41.5W	61	15	245	07m34s
0165	1608 Feb 16	01:03:27	97	124	T	-0.2154	1.0515	24.8S	171.6E	77	345	175	04m29s
0166	1608 Aug 10	15:00:05	96	129	A	0.2722	0.9581	31.0N	39.6W	74	194	158	04m46s
0167	1609 Feb 04	15:43:42	95	134	H	-0.9224	1.0041	77.3S	7.2E	22	297	37	00m15s
0168	1610 Jun 21	03:22:59	93	111	T	-0.8193	1.0705	31.5S	135.4E	35	354	400	05m59s
0169	1610 Dec 15	07:06:47	93	116	A	0.6195	0.9153	14.7N	78.2E	52	188	409	11m56s
0170	1611 Jun 10	20:34:25	92	121	T	-0.0836	1.0686	18.4N	127.7W	85	350	224	06m16s
0171	1611 Dec 04	08:03:42	91	126	A	-0.0804	0.9498	26.9S	55.9E	85	13	185	05m44s
0172	1612 May 30	10:34:28	91	131	T	0.6976	1.0135	63.6N	2.0W	45	149	65	00m58s
0173	1612 Nov 22	16:04:34	90	136	H	-0.7692	1.0002	65.7S	98.5W	39	43	1	00m01s
0174	1614 Apr 09	02:52:57	88	113	A	-0.6103	0.9411	25.7S	156.1E	52	329	268	06m22s
0175	1614 Oct 03	12:04:50	87	118	T	0.5511	1.0282	25.2N	13.5E	56	212	113	02m23s
0176	1615 Mar 29	07:03:24	87	123	A	0.1461	0.9851	10.7N	71.6E	82	151	53	01m28s
0177	1615 Sep 22	22:27:20	86	128	A	-0.1850	0.9784	9.1S	163.6W	79	29	78	02m11s
0178	1616 Mar 17	18:21:44	85	133	T	0.8568	1.0280	48.0N	131.4W	31	134	180	01m58s
0179	1616 Sep 11	01:44:05	85	138	A	-0.9506	0.9319	54.1S	102.3E	18	55	808	05m42s
0180	1618 Jan 26	13:46:43	83	115	A	-0.4700	0.9955	44.7S	9.8W	62	333	18	00m23s

Cat Num	Calendar Date	TD of Greatest Eclipse	ΔT s	Saros Num	Ecl Type	Gamma	Ecl Mag	Lat °	Long °	Sun Alt °	Sun Azm °	Path Width km	Central Line Dur
0181	1618 Jul 21	19:44:29	82	120	T	0.3558	1.0260	40.4N	106.4W	69	201	94	02m13s
0182	1619 Jan 15	20:38:06	81	125	A	0.2349	0.9422	8.1S	130.4W	76	165	220	07m16s
0183	1619 Jul 11	10:29:58	81	130	T	-0.4078	1.0718	1.3S	18.5E	66	12	255	06m41s
0184	1620 Jan 04	20:51:03	80	135	A	0.9322	0.9081	45.0N	146.5W	21	165	976	10m13s
0185	1621 May 21	08:53:43	78	112	A	0.6827	0.9962	63.1N	36.1E	47	167	18	00m18s
0186	1621 Nov 13	14:23:12	77	117	T	-0.6187	1.0194	55.8S	49.8W	52	16	84	01m28s
0187	1622 May 10	13:55:34	77	122	Am	-0.0758	0.9531	13.5N	28.8W	86	350	172	06m07s
0188	1622 Nov 03	05:34:47	76	127	T	0.0789	1.0544	10.7S	93.6E	86	193	180	05m01s
0189	1623 Apr 29	14:15:59	75	132	A	-0.8245	0.9378	39.8S	20.5W	34	344	405	06m54s
0190	1623 Oct 23	21:17:09	75	137	T	0.7770	1.0298	37.8N	128.1W	39	199	159	02m31s
0191	1625 Mar 08	17:32:38	73	114	T	0.4965	1.0434	23.9N	89.4W	60	161	166	03m50s
0192	1625 Sep 01	21:06:57	72	119	A	-0.5520	0.9380	24.2S	146.4W	56	18	274	07m37s
0193	1626 Feb 26	09:37:26	72	124	T	-0.1971	1.0535	19.7S	42.7E	79	343	180	04m42s
0194	1626 Aug 21	21:47:41	71	129	A	0.1976	0.9584	23.1N	142.9W	78	195	154	04m54s
0195	1627 Feb 16	00:13:30	70	134	H	-0.9119	1.0040	72.3S	131.0W	24	307	34	00m15s
0196	1627 Aug 11	04:17:13	70	139	H	0.9401	1.0001	77.7N	173.3W	19	253	1	00m00s
0197	1628 Jul 01	10:50:38	68	111	T	-0.8917	1.0692	40.3S	19.9E	27	358	501	05m32s
0198	1628 Dec 25	15:08:46	68	116	A	0.6264	0.9153	15.4N	44.1W	51	184	413	12m02s
0199	1629 Jun 21	03:59:23	67	121	T	-0.1580	1.0670	14.5N	121.6E	81	354	221	06m20s
0200	1629 Dec 14	16:19:07	66	126	A	-0.0726	0.9513	27.6S	66.3W	86	9	179	05m38s
0201	1630 Jun 10	17:41:06	66	131	H	0.6245	1.0122	60.9N	98.3W	51	161	54	00m55s
0202	1630 Dec 04	00:38:58	66	136	H	-0.7585	1.0017	68.7S	139.5E	40	36	9	00m07s
0203	1632 Apr 19	09:54:29	66	113	A	-0.6694	0.9447	26.4S	50.7E	48	331	267	06m03s
0204	1632 Oct 13	20:09:38	66	118	T	0.5872	1.0220	23.7N	108.2W	54	210	91	01m55s
0205	1633 Apr 08	14:37:05	66	123	A	0.0976	0.9913	12.5N	41.3W	84	152	31	00m51s
0206	1633 Oct 03	06:00:37	66	128	Am	-0.1405	0.9726	11.2S	83.4E	82	29	99	02m48s
0207	1634 Mar 29	02:25:11	66	133	T	0.8169	1.0346	48.7N	108.6E	35	133	198	02m24s
0208	1634 Sep 22	08:47:03	66	138	A	-0.8948	0.9300	51.5S	2.3E	26	51	572	06m03s
0209	1636 Feb 06	22:14:32	66	115	A	-0.4825	0.9943	41.6S	134.7W	61	329	23	00m29s
0210	1636 Aug 01	02:58:15	66	120	T	0.4279	1.0275	41.5N	148.9E	64	207	103	02m15s
0211	1637 Jan 26	04:48:31	66	125	A	0.2265	0.9412	6.4S	107.0E	77	161	223	07m12s
0212	1637 Jul 21	17:57:08	66	130	T	-0.3335	1.0731	1.8N	93.4W	71	16	251	06m37s
0213	1638 Jan 15	04:51:53	66	135	A	0.9242	0.9090	44.0N	88.9E	22	159	907	10m00s
0214	1639 Jun 01	15:55:15	66	112	A	0.7597	0.9930	71.7N	65.3W	40	173	38	00m31s
0215	1639 Nov 24	22:58:55	66	117	T	-0.6278	1.0197	59.6S	174.6W	51	10	87	01m27s
0216	1640 May 20	20:37:51	54	122	A	0.0002	0.9533	20.4N	130.3W	90	180	171	06m00s
0217	1640 Nov 13	14:11:19	54	127	T	0.0623	1.0522	14.8S	35.8W	87	189	173	04m50s
0218	1641 May 09	21:01:18	54	132	A	-0.7533	0.9425	30.8S	127.3W	41	349	321	06m56s
0219	1641 Nov 03	05:40:09	54	137	T	0.7570	1.0252	33.0N	102.5E	41	194	130	02m15s
0220	1643 Mar 20	01:47:18	54	114	T	0.5270	1.0479	30.0N	144.6E	58	159	186	04m02s
0221	1643 Sep 13	04:01:20	54	119	A	-0.6144	0.9343	32.3S	106.3E	52	21	307	07m35s
0222	1644 Mar 08	18:02:42	54	124	T	-0.1717	1.0555	14.0S	84.7W	80	342	186	04m57s
0223	1644 Sep 01	04:45:27	54	129	A	0.1307	0.9584	15.4N	110.9E	82	197	152	05m00s
0224	1645 Feb 26	08:35:06	54	134	H	-0.8956	1.0043	66.7S	94.3E	26	316	34	00m17s
0225	1645 Aug 21	11:34:17	54	139	H	0.8710	1.0040	68.2N	43.7E	29	222	28	00m16s
0226	1646 Jul 12	18:18:18	54	111	T	-0.9640	1.0658	53.1S	98.0W	15	5	834	04m44s
0227	1647 Jan 05	23:10:59	54	116	A	0.6336	0.9161	16.9N	166.5W	51	179	413	11m50s
0228	1647 Jul 02	11:21:20	54	121	T	-0.2344	1.0643	9.6N	11.0E	77	359	217	06m15s
0229	1647 Dec 26	00:38:34	54	126	A	-0.0675	0.9535	27.4S	170.6E	86	4	170	05m25s
0230	1648 Jun 21	00:43:22	54	131	H	0.5483	1.0102	56.7N	164.0E	56	171	42	00m49s
0231	1648 Dec 14	09:17:54	54	136	H	-0.7510	1.0035	70.9S	19.6E	41	25	18	00m14s
0232	1650 Apr 30	16:48:48	44	113	A	-0.7347	0.9481	28.5S	53.0W	43	334	273	05m43s
0233	1650 Oct 25	04:21:25	44	118	T	0.6170	1.0159	22.3N	127.8E	52	207	68	01m26s
0234	1651 Apr 19	22:04:36	44	123	A	0.0433	0.9976	13.7N	152.4W	87	154	8	00m14s
0235	1651 Oct 14	13:40:55	44	128	A	-0.1025	0.9668	13.5S	31.3W	84	28	120	03m27s
0236	1652 Apr 08	10:22:27	44	133	T	0.7713	1.0412	49.6N	8.9W	39	135	213	02m49s
0237	1652 Oct 02	15:58:29	44	138	A	-0.8459	0.9275	51.2S	102.8W	32	50	497	06m19s
0238	1654 Feb 17	06:36:37	44	115	A	-0.4991	0.9933	38.3S	100.9E	60	327	27	00m34s
0239	1654 Aug 12	10:17:43	44	120	T	0.4962	1.0285	41.7N	42.5E	60	211	110	02m16s
0240	1655 Feb 06	12:51:54	44	125	A	0.2130	0.9408	4.2S	14.0W	78	157	224	07m03s

Cat Num	Calendar Date	TD of Greatest Eclipse	ΔT s	Saros Num	Ecl Type	Gamma	Ecl Mag	Lat °	Long °	Sun Alt °	Sun Azm °	Path Width km	Central Line Dur
0241	1655 Aug 02	01:28:36	44	130	T	-0.2625	1.0735	3.7N	154.0E	75	20	247	06m28s
0242	1656 Jan 26	12:48:10	44	135	A	0.9122	0.9106	43.2N	34.0W	24	154	820	09m38s
0243	1656 Jul 21	18:39:47	44	140	T-	-0.9983	1.0244	63.4S	140.6W	0	39	-	-
0244	1657 Jun 11	22:52:09	44	112	A	0.8394	0.9888	80.5N	153.7W	33	190	73	00m45s
0245	1657 Dec 05	07:39:36	44	117	T	-0.6335	1.0205	62.1S	61.4E	50	3	91	01m29s
0246	1658 Jun 01	03:11:37	44	122	A	0.0828	0.9532	27.0N	131.3E	85	178	172	05m49s
0247	1658 Nov 24	22:54:41	44	127	T	0.0514	1.0502	18.0S	166.3W	87	186	167	04m40s
0248	1659 May 21	03:38:52	44	132	A	-0.6747	0.9469	22.2S	129.2E	47	353	264	06m50s
0249	1659 Nov 14	14:10:07	44	137	T	0.7432	1.0208	29.2N	28.2W	42	190	106	01m56s
0250	1661 Mar 30	09:55:23	35	114	T	0.5634	1.0524	36.7N	20.2E	55	158	209	04m12s
0251	1661 Sep 23	11:02:33	35	119	A	-0.6711	0.9306	40.3S	3.0W	48	23	347	07m29s
0252	1662 Mar 20	02:21:48	35	124	T	-0.1414	1.0576	7.9S	149.1E	82	342	192	05m11s
0253	1662 Sep 12	11:50:44	35	129	A	0.0694	0.9582	7.9N	2.6E	86	197	153	05m05s
0254	1663 Mar 09	16:48:41	35	134	H	-0.8735	1.0049	60.5S	37.1W	29	323	35	00m21s
0255	1663 Sep 01	18:59:08	35	139	H	0.8073	1.0065	58.6N	78.8W	36	212	38	00m29s
0256	1665 Jan 16	07:11:51	35	116	A	0.6419	0.9174	19.1N	71.2E	50	175	409	11m24s
0257	1665 Jul 12	18:44:05	35	121	T	-0.3095	1.0611	3.9N	100.6W	72	3	211	06m02s
0258	1666 Jan 05	08:58:50	35	126	A	-0.0624	0.9562	26.3S	47.1E	86	359	160	05m07s
0259	1666 Jul 02	07:42:29	35	131	H	0.4705	1.0075	51.4N	64.5E	62	178	29	00m39s
0260	1666 Dec 25	17:59:15	35	136	H	-0.7452	1.0058	71.6S	98.3W	42	11	30	00m24s
0261	1668 May 10	23:37:24	35	113	A	-0.8048	0.9510	32.3S	155.3W	36	336	296	05m21s
0262	1668 Nov 04	12:40:05	35	118	H	0.6401	1.0102	21.1N	1.9E	50	204	45	00m57s
0263	1669 Apr 30	05:26:06	35	123	H	-0.0170	1.0036	14.1N	98.2E	89	334	13	00m22s
0264	1669 Oct 24	21:28:04	35	128	A	-0.0710	0.9613	15.9S	147.7W	86	26	141	04m07s
0265	1670 Apr 19	18:12:20	28	133	T	0.7191	1.0476	50.6N	123.3W	44	137	225	03m15s
0266	1670 Oct 13	23:19:00	28	138	A	-0.8044	0.9247	52.4S	149.1E	36	49	467	06m34s
0267	1672 Feb 28	14:50:42	28	115	A	-0.5218	0.9926	35.2S	21.9W	58	326	30	00m38s
0268	1672 Aug 22	17:44:06	28	120	T	0.5593	1.0288	41.2N	66.2W	56	215	117	02m15s
0269	1673 Feb 16	20:49:18	28	125	A	0.1951	0.9409	1.8S	133.5W	79	154	223	06m52s
0270	1673 Aug 12	09:04:05	28	130	T	-0.1946	1.0731	4.6N	40.6E	79	23	242	06m15s
0271	1674 Feb 05	20:41:35	28	135	A	0.8979	0.9129	42.8N	155.7W	26	149	736	09m09s
0272	1674 Aug 02	02:07:57	28	140	T	-0.9295	1.0560	45.9S	120.9E	21	29	498	04m08s
0273	1675 Jun 23	05:44:38	28	112	A	0.9218	0.9835	84.1N	166.1W	22	282	154	01m01s
0274	1675 Dec 16	16:24:03	28	117	T	-0.6367	1.0218	63.1S	62.0W	50	353	97	01m33s
0275	1676 Jun 11	09:42:36	28	122	A	0.1672	0.9527	33.0N	34.6E	80	182	176	05m38s
0276	1676 Dec 05	07:42:08	28	127	T	0.0436	1.0486	20.2S	62.5E	88	181	162	04m30s
0277	1677 May 31	10:13:53	28	132	A	-0.5935	0.9510	14.4S	27.5E	53	358	223	06m35s
0278	1677 Nov 24	22:44:03	28	137	T	0.7332	1.0166	26.3N	159.6W	43	186	84	01m36s
0279	1679 Apr 10	17:55:13	28	114	T	0.6070	1.0565	43.8N	102.2W	52	157	233	04m17s
0280	1679 Oct 04	18:13:56	28	119	A	-0.7191	0.9270	48.0S	114.8W	44	26	391	07m21s
0281	1680 Mar 30	10:32:00	22	124	T	-0.1039	1.0595	1.5S	24.9E	84	343	197	05m25s
0282	1680 Sep 22	19:06:23	22	129	A	0.0161	0.9578	0.7N	108.2W	89	198	153	05m08s
0283	1681 Mar 20	00:52:58	22	134	H	-0.8445	1.0057	53.8S	165.2W	32	329	37	00m26s
0284	1681 Sep 12	02:33:12	22	139	H	0.7504	1.0083	49.8N	161.2E	41	207	43	00m40s
0285	1683 Jan 27	15:10:08	22	116	A	0.6525	0.9195	22.1N	50.5W	49	171	401	10m44s
0286	1683 Jul 24	02:06:59	22	121	T	-0.3838	1.0569	2.5S	147.1E	67	7	203	05m38s
0287	1684 Jan 16	17:18:52	22	126	A	-0.0565	0.9597	24.2S	76.6W	87	355	147	04m43s
0288	1684 Jul 12	14:40:34	22	131	H	0.3927	1.0041	45.2N	37.1W	67	184	16	00m23s
0289	1685 Jan 05	02:42:50	22	136	H	-0.7409	1.0086	70.7S	143.2E	42	357	44	00m35s
0290	1686 May 22	06:21:19	22	113	A	-0.8791	0.9533	38.6S	103.4E	28	339	353	04m56s
0291	1686 Nov 15	21:04:59	22	118	H	0.6578	1.0048	20.2N	126.0W	49	200	22	00m28s
0292	1687 May 11	12:42:28	22	123	H	-0.0827	1.0094	13.6N	9.8W	85	339	33	00m57s
0293	1687 Nov 05	05:22:24	22	128	A	-0.0461	0.9561	18.3S	94.4E	87	23	160	04m49s
0294	1688 Apr 30	01:57:33	22	133	T	0.6621	1.0535	51.4N	124.5E	48	141	234	03m40s
0295	1688 Oct 24	06:46:41	22	138	A	-0.7686	0.9221	54.4S	39.3E	39	47	453	06m49s
0296	1690 Mar 10	22:56:00	17	115	A	-0.5511	0.9920	32.5S	142.9W	56	325	33	00m42s
0297	1690 Sep 03	01:17:46	17	120	T	0.6173	1.0287	40.3N	177.4W	52	217	122	02m13s
0298	1691 Feb 28	04:37:41	17	125	A	0.1701	0.9414	0.8N	109.3E	80	152	220	06m40s
0299	1691 Aug 23	16:45:57	17	130	T	-0.1317	1.0720	4.5N	74.3W	82	26	236	06m01s
0300	1692 Feb 17	04:26:55	17	135	A	0.8765	0.9159	42.4N	85.6E	28	145	644	08m36s

Cat Num	Calendar Date	TD of Greatest Eclipse	ΔT s	Saros Num	Ecl Type	Gamma	Ecl Mag	Lat °	Long °	Sun Alt °	Sun Azm °	Path Width km	Central Line Dur
0301	1692 Aug 12	09:41:05	17	140	T	-0.8649	1.0546	39.8S	8.6E	30	31	353	04m10s
0302	1693 Dec 27	01:10:50	17	117	T	-0.6387	1.0236	62.5S	174.3E	50	343	105	01m39s
0303	1694 Jun 22	16:08:45	17	122	A	0.2556	0.9517	38.4N	59.7W	75	187	183	05m27s
0304	1694 Dec 16	16:33:11	17	127	T	0.0389	1.0475	21.3S	69.2W	88	176	158	04m22s
0305	1695 Jun 11	16:44:23	17	132	A	-0.5077	0.9545	7.4S	72.2W	59	2	193	06m13s
0306	1695 Dec 06	07:23:18	17	137	T	0.7280	1.0128	24.3N	68.0E	43	181	64	01m16s
0307	1697 Apr 21	01:49:21	17	114	T	0.6559	1.0602	51.4N	136.9E	49	157	262	04m18s
0308	1697 Oct 15	01:33:41	17	119	A	-0.7603	0.9236	55.4S	131.3E	40	29	441	07m12s
0309	1698 Apr 10	18:34:25	17	124	Tm	-0.0600	1.0613	5.1N	97.3W	87	344	201	05m36s
0310	1698 Oct 04	02:31:24	17	129	A	-0.0305	0.9573	6.2S	138.9E	88	17	155	05m10s
0311	1699 Mar 31	08:48:44	17	134	H	-0.8089	1.0065	46.8S	69.7E	36	334	38	00m32s
0312	1699 Sep 23	10:16:12	17	139	H	0.6999	1.0095	41.8N	40.7E	45	204	46	00m49s
0313	1701 Feb 07	23:04:53	14	116	A	0.6662	0.9219	25.9N	171.6W	48	167	393	09m55s
0314	1701 Aug 04	09:31:44	14	121	T	-0.4558	1.0521	9.4S	33.7E	63	10	193	05m06s
0315	1702 Jan 28	01:37:09	14	126	A	-0.0485	0.9636	21.2S	159.6E	87	351	132	04m14s
0316	1702 Jul 24	21:38:51	14	131	H	0.3160	1.0002	38.4N	140.4W	71	188	1	00m01s
0317	1703 Jan 17	11:24:25	14	136	H2	-0.7345	1.0120	67.9S	22.2E	42	347	61	00m50s
0318	1704 Jun 02	13:02:36	14	113	A	-0.9561	0.9542	49.1S	3.4E	16	341	578	04m26s
0319	1704 Nov 27	05:33:52	14	118	A	0.6716	0.9999	19.7N	105.0E	48	196	1	00m01s
0320	1705 May 22	19:55:06	14	123	Hm	-0.1524	1.0147	12.2N	117.0W	81	343	51	01m32s
0321	1705 Nov 16	13:23:06	14	128	A	-0.0271	0.9514	20.4S	25.0W	88	19	178	05m31s
0322	1706 May 12	09:35:08	14	133	T	0.5984	1.0591	51.5N	15.2E	53	147	242	04m06s
0323	1706 Nov 05	14:23:56	14	138	A	-0.7407	0.9196	57.0S	72.6W	42	44	449	07m02s
0324	1708 Mar 22	06:51:36	14	115	A	-0.5878	0.9913	30.4S	98.3E	54	326	37	00m46s
0325	1708 Sep 14	09:00:22	14	120	T	0.6685	1.0281	39.2N	68.3E	48	218	126	02m10s
0326	1709 Mar 11	12:18:34	14	125	Am	0.1395	0.9422	3.4N	5.9W	82	151	216	06m29s
0327	1709 Sep 04	00:32:26	14	130	T	-0.0725	1.0703	3.7N	169.7E	86	28	229	05m47s
0328	1710 Feb 28	12:07:28	12	135	A	0.8509	0.9194	42.5N	31.2W	31	141	562	08m00s
0329	1710 Aug 24	17:17:15	12	140	T	-0.8032	1.0519	36.5S	105.1W	36	33	282	04m00s
0330	1712 Jan 08	09:58:39	12	117	T	-0.6405	1.0258	60.6S	49.2E	50	335	114	01m48s
0331	1712 Jul 03	22:34:57	12	122	A	0.3433	0.9503	42.8N	152.7W	70	194	194	05m18s
0332	1712 Dec 28	01:24:54	12	127	T	0.0347	1.0466	21.5S	159.0E	88	171	155	04m15s
0333	1713 Jun 22	23:15:38	12	132	A	-0.4217	0.9576	1.3S	171.1W	65	6	170	05m45s
0334	1713 Dec 17	16:04:20	12	137	H	0.7249	1.0094	23.1N	64.8W	43	176	47	00m56s
0335	1715 May 03	09:36:30	12	114	T	0.7112	1.0632	59.4N	17.9E	44	157	295	04m14s
0336	1715 Oct 27	09:02:48	12	119	A	-0.7939	0.9206	62.5S	15.5E	37	31	494	07m02s
0337	1716 Apr 22	02:28:32	12	124	T	-0.0091	1.0625	11.8N	142.6E	90	343	205	05m43s
0338	1716 Oct 15	10:07:38	12	129	A	-0.0687	0.9570	12.5S	23.5E	86	16	157	05m10s
0339	1717 Apr 11	16:34:40	12	134	H	-0.7660	1.0072	39.5S	52.1W	40	339	39	00m39s
0340	1717 Oct 04	18:08:27	12	139	H	0.6563	1.0104	34.6N	81.1W	49	201	47	00m56s
0341	1719 Feb 19	06:52:57	12	116	A	0.6855	0.9250	30.5N	68.6E	47	163	384	09m01s
0342	1719 Aug 15	16:59:51	12	121	T	-0.5243	1.0466	16.8S	81.1W	58	13	181	04m27s
0343	1720 Feb 08	09:52:31	12	126	A	-0.0375	0.9681	17.4S	36.1E	88	348	115	03m40s
0344	1720 Aug 04	04:38:14	12	131	A	0.2410	0.9957	31.1N	114.8E	76	192	16	00m27s
0345	1721 Jan 27	20:05:11	12	136	T	-0.7269	1.0158	64.0S	102.4W	43	340	79	01m07s
0346	1722 Dec 08	14:07:34	12	118	A	0.6808	0.9955	19.5N	25.4W	47	191	21	00m28s
0347	1723 Jun 03	03:05:13	12	123	T	-0.2250	1.0196	9.6N	136.1E	77	347	69	02m05s
0348	1723 Nov 27	21:28:15	12	128	A	-0.0125	0.9471	22.0S	145.2W	89	14	195	06m12s
0349	1724 May 22	17:10:08	12	133	T	0.5319	1.0640	50.8N	92.9W	58	154	247	04m33s
0350	1724 Nov 15	22:07:37	12	138	A	-0.7184	0.9174	59.9S	175.0E	44	40	448	07m15s
0351	1726 Apr 02	14:38:16	12	115	A	-0.6312	0.9906	29.2S	18.3W	51	327	42	00m52s
0352	1726 Sep 25	16:51:44	12	120	T	0.7133	1.0273	38.0N	48.9W	44	218	129	02m07s
0353	1727 Mar 22	19:47:55	12	125	A	0.0996	0.9432	5.7N	118.0W	84	151	211	06m20s
0354	1727 Sep 15	08:27:30	12	130	T	-0.0202	1.0681	2.2N	51.4E	89	29	222	05m33s
0355	1728 Mar 10	19:38:55	12	135	A	0.8172	0.9233	42.8N	144.6W	35	139	485	07m25s
0356	1728 Sep 04	00:59:22	12	140	T	-0.7466	1.0484	35.0S	139.6E	41	34	236	03m44s
0357	1730 Jan 18	18:45:15	13	117	T	-0.6439	1.0286	57.8S	77.4W	50	329	126	01m59s
0358	1730 Jul 15	04:59:08	13	122	A	0.4325	0.9484	46.3N	116.0E	64	200	210	05m13s
0359	1731 Jan 08	10:17:44	13	127	Tm	0.0313	1.0464	20.7S	27.0E	88	166	155	04m10s
0360	1731 Jul 04	05:46:25	13	132	A	-0.3341	0.9602	3.8N	90.8E	71	10	153	05m15s

Cat Num	Calendar Date	TD of Greatest Eclipse	ΔT s	Saros Num	Ecl Type	Gamma	Ecl Mag	Lat °	Long °	Sun Alt °	Sun Azm °	Path Width km	Central Line Dur
0361	1731 Dec 29	00:46:52	13	137	H	0.7234	1.0065	22.7N	162.2E	44	171	32	00m39s
0362	1733 May 13	17:18:28	13	114	T	0.7712	1.0656	67.9N	99.5W	39	157	339	04m06s
0363	1733 Nov 06	16:40:14	13	119	A	-0.8208	0.9179	69.0S	101.2W	34	32	548	06m53s
0364	1734 May 03	10:15:56	13	124	T	0.0472	1.0635	18.4N	24.6E	87	168	208	05m46s
0365	1734 Oct 26	17:53:27	13	129	A	-0.0996	0.9567	18.2S	93.8W	84	14	159	05m08s
0366	1735 Apr 23	00:11:35	13	134	H	-0.7164	1.0077	32.2S	171.0W	44	343	38	00m44s
0367	1735 Oct 16	02:10:33	13	139	H	0.6202	1.0110	28.3N	155.2E	51	198	48	01m02s
0368	1737 Mar 01	14:35:17	13	116	A	0.7098	0.9283	36.0N	50.1W	45	160	378	08m04s
0369	1737 Aug 26	00:32:08	13	121	T	-0.5885	1.0407	24.4S	162.5E	54	17	167	03m44s
0370	1738 Feb 18	18:02:31	13	126	A	-0.0211	0.9732	12.8S	86.7W	89	346	96	03m03s
0371	1738 Aug 15	11:40:12	13	131	A	0.1688	0.9907	23.7N	8.4E	80	194	33	01m00s
0372	1739 Feb 08	04:41:13	13	136	T	-0.7150	1.0203	59.2S	131.0E	44	336	99	01m27s
0373	1739 Aug 04	15:40:56	13	141	A	0.9588	0.9408	79.9N	42.9E	16	280	801	03m59s
0374	1740 Dec 18	22:43:17	15	118	A	0.6876	0.9917	19.9N	156.4W	46	187	40	00m53s
0375	1741 Jun 13	10:12:47	15	123	T	-0.3007	1.0239	6.0N	29.4E	73	352	85	02m35s
0376	1741 Dec 08	05:37:59	15	128	A	-0.0024	0.9434	23.0S	93.6E	90	6	209	06m51s
0377	1742 Jun 03	00:39:56	15	133	T	0.4607	1.0683	49.0N	160.2E	62	161	251	05m00s
0378	1742 Nov 27	05:58:59	15	138	A	-0.7019	0.9156	62.6S	62.2E	45	34	450	07m26s
0379	1744 Apr 12	22:15:24	15	115	A	-0.6819	0.9895	29.0S	132.6W	47	329	49	00m59s
0380	1744 Oct 06	00:51:23	15	120	T	0.7521	1.0263	37.0N	169.1W	41	216	132	02m04s
0381	1745 Apr 02	03:09:18	15	125	A	0.0536	0.9444	7.7N	132.2E	87	152	205	06m13s
0382	1745 Sep 25	16:28:55	15	130	Tm	0.0269	1.0655	0.3N	68.6W	88	209	214	05m21s
0383	1746 Mar 22	03:02:48	15	135	A	0.7771	0.9277	43.5N	104.7E	39	138	419	06m51s
0384	1746 Sep 15	08:46:36	15	140	T	-0.6948	1.0441	34.9S	23.1E	46	36	201	03m23s
0385	1748 Jan 30	03:29:13	15	117	T	-0.6501	1.0317	54.4S	154.8E	49	324	140	02m12s
0386	1748 Jul 25	11:27:02	15	122	A	0.5183	0.9461	48.7N	24.5E	59	207	231	05m12s
0387	1749 Jan 18	19:08:56	15	127	T	0.0265	1.0465	19.1S	104.9W	89	161	155	04m07s
0388	1749 Jul 14	12:19:20	15	132	A	-0.2476	0.9623	7.8N	7.2W	76	14	141	04m46s
0389	1750 Jan 08	09:28:43	17	137	H	0.7217	1.0041	23.0N	29.3E	44	167	20	00m24s
0390	1751 May 25	00:55:15	17	114	T	0.8359	1.0670	77.0N	144.7E	33	157	402	03m53s
0391	1751 Nov 18	00:26:00	17	119	A	-0.8411	0.9159	74.9S	142.8E	32	31	597	06m45s
0392	1752 May 13	17:56:28	17	124	T	0.1090	1.0637	24.9N	91.1W	84	171	210	05m42s
0393	1752 Nov 06	01:48:14	17	129	A	-0.1239	0.9567	23.2S	147.4E	83	12	159	05m03s
0394	1753 May 03	07:39:40	17	134	H	-0.6601	1.0079	24.9S	73.0E	49	347	36	00m48s
0395	1753 Oct 26	10:22:00	17	139	H	0.5911	1.0115	22.7N	29.7E	54	195	49	01m08s
0396	1755 Mar 12	22:09:32	17	116	A	0.7413	0.9319	42.2N	167.4W	42	156	375	07m07s
0397	1755 Sep 06	08:09:45	17	121	T	-0.6478	1.0342	32.1S	44.3E	49	20	150	03m00s
0398	1756 Mar 01	02:07:09	17	126	A	0.0006	0.9787	7.5S	151.4E	90	29	76	02m24s
0399	1756 Aug 25	18:46:17	17	131	Am	0.1009	0.9853	16.1N	99.5W	84	196	52	01m38s
0400	1757 Feb 18	13:14:12	17	136	T	-0.7000	1.0251	53.8S	2.9E	45	335	119	01m51s
0401	1757 Aug 14	22:16:45	17	141	A	0.8808	0.9407	71.6N	113.4W	28	224	467	04m36s
0402	1758 Dec 30	07:20:12	17	118	A	0.6929	0.9885	20.8N	72.2E	46	182	56	01m15s
0403	1759 Jun 24	17:20:59	17	123	T	-0.3767	1.0275	1.4N	78.0W	68	356	101	02m59s
0404	1759 Dec 19	13:50:05	17	128	A	0.0051	0.9404	23.3S	28.0W	90	193	221	07m25s
0405	1760 Jun 13	08:09:15	19	133	T	0.3884	1.0719	46.0N	52.7E	67	168	254	05m27s
0406	1760 Dec 07	13:53:44	19	138	A	-0.6882	0.9144	64.7S	49.4W	46	26	451	07m36s
0407	1762 Apr 24	05:42:10	19	115	A	-0.7402	0.9881	30.3S	115.7E	42	331	61	01m08s
0408	1762 Oct 17	09:00:34	19	120	T	0.7836	1.0253	36.2N	67.6E	38	214	135	02m02s
0409	1763 Apr 13	10:19:31	19	125	A	-0.0010	0.9455	9.0N	25.4E	90	249	201	06m11s
0410	1763 Oct 07	00:39:04	19	130	T	0.0666	1.0627	2.0S	169.1E	86	209	206	05m09s
0411	1764 Apr 01	10:17:15	19	135	A	0.7289	0.9323	44.2N	2.5W	43	138	361	06m20s
0412	1764 Sep 25	16:41:43	19	140	T	-0.6502	1.0394	36.0S	95.4W	49	37	171	03m01s
0413	1766 Feb 09	12:09:44	19	117	T	-0.6597	1.0352	50.7S	26.6E	48	321	156	02m27s
0414	1766 Aug 05	17:56:57	19	122	A	0.6023	0.9433	50.2N	66.9W	53	214	260	05m15s
0415	1767 Jan 30	03:56:55	19	127	T	0.0190	1.0471	16.8S	123.9E	89	157	157	04m06s
0416	1767 Jul 25	18:55:47	19	132	A	-0.1630	0.9638	10.8N	105.4W	81	18	132	04m21s
0417	1768 Jan 19	18:09:29	19	137	H	0.7195	1.0022	23.9N	103.2W	44	162	11	00m13s
0418	1768 Jul 14	01:40:57	19	142	H	-0.9176	1.0055	43.0S	137.4E	23	19	48	00m29s
0419	1769 Jun 04	08:28:34	19	114	T	0.9037	1.0671	87.3N	26.2E	25	153	521	03m36s
0420	1769 Nov 28	08:18:40	19	119	A	-0.8559	0.9144	80.0S	32.0E	31	22	638	06m38s

Cat Num	Calendar Date	TD of Greatest Eclipse	ΔT s	Saros Num	Ecl Type	Gamma	Ecl Mag	Lat °	Long °	Sun Alt °	Sun Azm °	Path Width km	Central Line Dur
0421	1770 May 25	01:30:11	21	124	T	0.1760	1.0634	31.2N	155.6E	80	174	211	05m31s
0422	1770 Nov 17	09:51:53	21	129	A	-0.1416	0.9571	27.3S	27.1E	82	9	158	04m56s
0423	1771 May 14	15:00:02	21	134	H	-0.5981	1.0076	17.8S	40.4W	53	351	33	00m49s
0424	1771 Nov 06	18:41:02	21	139	H	0.5676	1.0120	17.9N	97.3W	55	192	50	01m13s
0425	1773 Mar 23	05:36:57	21	116	A	0.7785	0.9357	49.2N	76.2E	39	152	378	06m13s
0426	1773 Sep 16	15:52:23	21	121	T	-0.7020	1.0275	39.9S	75.5W	45	23	130	02m18s
0427	1774 Mar 12	10:05:14	21	126	A	0.0284	0.9845	1.7S	30.8E	88	162	55	01m43s
0428	1774 Sep 06	01:57:40	21	131	A	0.0385	0.9797	8.7N	150.9E	88	197	72	02m20s
0429	1775 Mar 01	21:39:20	21	136	T	-0.6783	1.0304	47.9S	124.7W	47	335	139	02m20s
0430	1775 Aug 26	04:59:40	21	141	A	0.8088	0.9391	61.3N	132.0E	36	213	383	05m16s
0431	1777 Jan 09	15:55:35	21	118	A	0.6987	0.9859	22.4N	58.9W	46	177	70	01m32s
0432	1777 Jul 05	00:29:29	21	123	T	-0.4531	1.0305	4.2S	173.7E	63	0	115	03m17s
0433	1777 Dec 29	22:03:28	21	128	A	0.0109	0.9380	22.7S	150.0W	90	183	231	07m53s
0434	1778 Jun 24	15:34:55	21	133	T	0.3127	1.0746	41.8N	55.0W	72	175	255	05m52s
0435	1778 Dec 18	21:53:53	21	138	A	-0.6788	0.9137	65.8S	160.6W	47	16	450	07m44s
0436	1780 May 04	13:00:41	21	115	A	-0.8043	0.9861	33.3S	6.0E	36	334	81	01m21s
0437	1780 Oct 27	17:18:27	21	120	T	0.8083	1.0244	35.6N	58.6W	36	210	138	02m00s
0438	1781 Apr 23	17:21:25	21	125	A	-0.0620	0.9467	9.7N	79.2W	87	334	197	06m13s
0439	1781 Oct 17	08:55:58	21	130	T	0.1007	1.0596	4.3S	45.1E	84	207	197	04m59s
0440	1782 Apr 12	17:24:47	21	135	A	0.6745	0.9370	45.1N	107.1W	47	140	311	05m51s
0441	1782 Oct 07	00:43:19	21	140	T	-0.6113	1.0344	37.9S	144.6E	52	37	144	02m37s
0442	1784 Feb 20	20:45:38	21	117	T	-0.6739	1.0389	47.2S	101.5W	47	320	174	02m44s
0443	1784 Aug 16	00:31:53	21	122	A	0.6819	0.9402	50.9N	159.8W	47	220	299	05m23s
0444	1785 Feb 09	12:40:41	21	127	T	0.0080	1.0480	14.1S	6.6W	90	150	159	04m07s
0445	1785 Aug 05	01:37:22	21	132	A	-0.0817	0.9650	12.7N	155.3E	85	22	127	04m01s
0446	1786 Jan 30	02:45:26	21	137	H	0.7141	1.0009	25.1N	125.5E	44	158	5	00m05s
0447	1786 Jul 25	08:46:33	21	142	T	-0.8384	1.0106	34.6S	30.9E	33	21	66	00m59s
0448	1787 Jun 15	15:59:25	21	114	T	0.9738	1.0648	78.7N	104.8E	12	346	997	03m09s
0449	1787 Dec 09	16:15:38	21	119	A	-0.8674	0.9136	83.4S	62.7W	29	357	672	06m32s
0450	1788 Jun 04	08:59:30	21	124	T	0.2465	1.0623	37.0N	44.4E	76	179	211	05m15s
0451	1788 Nov 27	18:02:53	21	129	A	-0.1542	0.9579	30.4S	94.3W	81	4	155	04m46s
0452	1789 May 24	22:11:58	21	134	H	-0.5297	1.0068	11.0S	151.0W	58	355	28	00m46s
0453	1789 Nov 17	03:08:34	21	139	H	0.5505	1.0126	14.1N	133.9E	57	188	52	01m19s
0454	1791 Apr 03	12:55:13	21	116	A	0.8236	0.9394	57.1N	39.5W	34	147	394	05m21s
0455	1791 Sep 27	23:42:29	21	121	T	-0.7492	1.0206	47.6S	162.4E	41	27	106	01m38s
0456	1792 Mar 22	17:57:34	21	126	A	0.0618	0.9906	4.5N	88.7W	86	162	33	01m02s
0457	1792 Sep 16	09:13:51	21	131	A	-0.0191	0.9739	1.3N	39.9E	89	18	93	03m02s
0458	1793 Mar 12	06:00:07	21	136	T	-0.6524	1.0359	41.7S	107.8E	49	336	158	02m51s
0459	1793 Sep 05	11:47:24	21	141	A	0.7408	0.9370	51.7N	23.0E	42	207	346	06m02s
0460	1795 Jan 21	00:29:13	21	118	A	0.7055	0.9837	24.8N	170.3E	45	173	81	01m44s
0461	1795 Jul 16	07:41:36	21	123	T	-0.5273	1.0327	10.4S	63.8E	58	4	130	03m26s
0462	1796 Jan 10	06:14:52	21	128	A	0.0179	0.9362	21.1S	88.3E	89	177	238	08m15s
0463	1796 Jul 04	23:02:54	21	133	T	0.2385	1.0764	36.8N	164.6W	76	180	255	06m15s
0464	1796 Dec 29	05:54:58	21	138	A	-0.6703	0.9136	65.5S	88.6E	48	5	446	07m51s
0465	1797 Jun 24	16:18:13	21	143	T	0.9780	1.0570	77.2N	134.0E	11	17	975	02m47s
0466	1798 May 15	20:10:32	21	115	A	-0.8744	0.9832	38.6S	101.5W	29	336	121	01m36s
0467	1798 Nov 08	01:44:38	21	120	T	0.8270	1.0237	35.1N	172.5E	34	206	141	01m59s
0468	1799 May 05	00:13:07	21	125	A	-0.1310	0.9476	9.3N	178.9E	83	338	194	06m20s
0469	1799 Oct 28	17:21:46	21	130	T	0.1274	1.0566	6.7S	81.2W	83	205	188	04m50s
0470	1800 Apr 24	00:24:00	18	135	A	0.6125	0.9417	45.7N	151.4E	52	143	269	05m27s
0471	1800 Oct 18	08:51:52	18	140	T	-0.5788	1.0293	40.3S	23.2E	54	36	120	02m14s
0472	1802 Mar 04	05:14:29	18	117	T	-0.6942	1.0428	44.0S	131.6E	46	320	196	03m02s
0473	1802 Aug 28	07:12:00	18	122	A	0.7568	0.9367	51.3N	105.7E	41	225	353	05m35s
0474	1803 Feb 21	21:18:46	18	127	T	-0.0075	1.0492	11.1S	135.9W	90	337	163	04m09s
0475	1803 Aug 17	08:25:03	18	132	A	-0.0048	0.9657	13.6N	54.7E	90	35	124	03m47s
0476	1804 Feb 11	11:16:32	18	137	H	0.7053	1.0001	26.7N	4.4W	45	153	0	00m00s
0477	1804 Aug 05	15:57:13	18	142	T	-0.7623	1.0144	29.3S	77.0W	40	24	75	01m20s
0478	1805 Dec 21	00:17:37	18	119	A	-0.8751	0.9134	83.1S	143.8W	29	317	692	06m26s
0479	1806 Jun 16	16:24:27	18	124	T	0.3203	1.0604	42.2N	64.5W	71	184	210	04m55s
0480	1806 Dec 10	02:19:39	18	129	A	-0.1627	0.9591	32.4S	143.4E	80	360	151	04m32s

Cat Num	Calendar Date	TD of Greatest Eclipse	ΔT s	Saros Num	Ecl Type	Gamma	Ecl Mag	Lat °	Long °	Sun Alt °	Sun Azm °	Path Width km	Central Line Dur
0481	1807 Jun 06	05:18:31	18	134	H	-0.4577	1.0055	4.7S	100.4E	63	359	21	00m38s
0482	1807 Nov 29	11:42:09	18	139	H	0.5377	1.0135	11.1N	3.9E	57	184	55	01m26s
0483	1809 Apr 14	20:07:11	18	116	A	0.8742	0.9429	65.8N	157.2W	29	139	435	04m35s
0484	1809 Oct 09	07:38:42	18	121	T	-0.7904	1.0137	55.1S	38.4E	37	30	77	01m02s
0485	1810 Apr 04	01:41:19	16	126	A	0.1031	0.9967	11.1N	153.8E	84	163	12	00m21s
0486	1810 Sep 28	16:37:25	16	131	A	-0.0696	0.9681	5.8S	72.8W	86	18	115	03m45s
0487	1811 Mar 24	14:12:13	16	136	T	-0.6190	1.0416	35.2S	18.0W	52	338	176	03m27s
0488	1811 Sep 17	18:43:45	16	141	A	0.6798	0.9345	43.0N	85.9W	47	204	330	06m51s
0489	1813 Feb 01	08:58:27	16	118	A	0.7151	0.9820	27.9N	40.4E	44	169	91	01m52s
0490	1813 Jul 27	14:55:35	16	123	T	-0.6006	1.0341	17.4S	47.4W	53	8	144	03m27s
0491	1814 Jan 21	14:24:46	16	128	A	0.0253	0.9350	18.6S	33.4W	89	173	242	08m28s
0492	1814 Jul 17	06:30:29	16	133	T	0.1641	1.0774	30.9N	84.7E	80	185	254	06m33s
0493	1815 Jan 10	13:57:06	16	138	A	-0.6626	0.9143	63.8S	23.5W	48	355	438	07m55s
0494	1815 Jul 06	23:43:07	16	143	T	0.9062	1.0593	88.1N	162.7W	25	192	470	03m13s
0495	1816 May 27	03:13:24	16	115	A	-0.9492	0.9791	48.0S	153.6E	18	338	238	01m54s
0496	1816 Nov 19	10:17:22	16	120	T	0.8407	1.0233	35.0N	41.5E	33	202	144	02m00s
0497	1817 May 16	06:58:14	16	125	A	-0.2049	0.9483	7.9N	78.5E	78	341	194	06m30s
0498	1817 Nov 09	01:53:53	16	130	T	0.1487	1.0536	8.9S	150.9E	82	202	179	04m42s
0499	1818 May 05	07:15:49	16	135	A	0.5440	0.9464	45.8N	52.5E	57	148	233	05m05s
0500	1818 Oct 29	17:07:09	16	140	T	-0.5524	1.0241	43.1S	99.4W	56	34	98	01m51s
0501	1820 Mar 14	13:37:15	17	117	T	-0.7199	1.0467	41.5S	5.8E	44	320	220	03m20s
0502	1820 Sep 07	13:59:58	17	122	A	0.8251	0.9330	51.6N	8.7E	34	229	432	05m49s
0503	1821 Mar 04	05:50:13	17	127	T	-0.0284	1.0506	8.0S	96.3E	88	333	168	04m14s
0504	1821 Aug 27	15:19:42	17	132	A	0.0671	0.9661	13.6N	47.7W	86	207	123	03m38s
0505	1822 Feb 21	19:40:40	17	137	A	0.6914	0.9996	28.6N	132.3W	46	150	2	00m02s
0506	1822 Aug 16	23:14:34	16	142	T	-0.6904	1.0173	26.1S	173.6E	46	27	80	01m35s
0507	1824 Jan 01	08:21:09	16	119	A	-0.8821	0.9139	79.9S	116.2E	28	295	705	06m21s
0508	1824 Jun 26	23:46:33	16	124	T	0.3960	1.0578	46.6N	171.4W	66	190	207	04m31s
0509	1824 Dec 20	10:40:35	16	129	Am	-0.1685	0.9610	33.3S	20.4E	80	354	144	04m15s
0510	1825 Jun 16	12:19:03	16	134	H	-0.3813	1.0036	1.0N	6.0W	68	3	13	00m25s
0511	1825 Dec 09	20:21:45	16	139	H2	0.5296	1.0148	9.2N	127.4W	58	180	60	01m34s
0512	1827 Apr 26	03:11:14	16	116	A	0.9316	0.9458	74.8N	73.5E	21	118	559	03m53s
0513	1827 Oct 20	15:42:05	16	121	H	-0.8251	1.0070	62.3S	87.6W	34	34	43	00m30s
0514	1828 Apr 14	09:19:38	16	126	Hm	0.1498	1.0029	17.9N	37.8E	81	164	10	00m18s
0515	1828 Oct 09	00:07:46	16	131	A	-0.1139	0.9623	12.5S	173.0E	83	17	137	04m26s
0516	1829 Apr 03	22:18:36	16	136	T	-0.5803	1.0474	28.5S	142.5W	54	341	192	04m05s
0517	1829 Sep 28	01:46:53	16	141	A	0.6244	0.9318	34.9N	164.4E	51	202	323	07m43s
0518	1831 Feb 12	17:21:44	11	118	A	0.7288	0.9807	31.9N	88.3W	43	165	100	01m57s
0519	1831 Aug 07	22:15:59	11	123	T	-0.6691	1.0349	24.9S	160.9W	48	12	158	03m20s
0520	1832 Feb 01	22:30:14	11	128	A	0.0354	0.9344	15.3S	154.4W	88	169	245	08m35s
0521	1832 Jul 27	14:01:06	11	133	T	0.0919	1.0776	24.5N	27.9W	85	188	252	06m46s
0522	1833 Jan 20	21:56:55	11	138	A	-0.6530	0.9155	60.6S	137.4W	49	347	426	07m59s
0523	1833 Jul 17	07:08:02	11	143	T	0.8348	1.0591	77.5N	92.5E	33	200	357	03m29s
0524	1834 Nov 30	18:56:35	11	120	T	0.8497	1.0233	34.9N	91.6W	32	197	150	02m02s
0525	1835 May 27	13:35:42	11	125	A	-0.2846	0.9486	5.3N	20.2W	74	345	196	06m44s
0526	1835 Nov 20	10:31:58	11	130	T	0.1649	1.0510	10.7S	21.6E	81	198	171	04m35s
0527	1836 May 15	14:01:39	11	135	A	0.4700	0.9509	45.1N	44.4W	62	153	203	04m47s
0528	1836 Nov 09	01:29:25	11	140	T	-0.5327	1.0191	46.1S	136.8E	58	31	77	01m28s
0529	1838 Mar 25	21:52:16	11	117	T	-0.7524	1.0505	39.7S	118.3W	41	321	249	03m39s
0530	1838 Sep 18	20:55:56	11	122	A	0.8868	0.9289	52.4N	90.6W	27	232	562	06m06s
0531	1839 Mar 15	14:13:42	11	127	T	-0.0557	1.0520	5.1S	29.5W	87	331	172	04m20s
0532	1839 Sep 07	22:23:26	11	132	Am	0.1324	0.9661	12.8N	152.6W	82	209	123	03m34s
0533	1840 Mar 04	03:58:22	8	137	A	0.6728	0.9995	30.6N	101.8E	48	147	2	00m03s
0534	1840 Aug 27	06:37:32	8	142	T	-0.6224	1.0195	24.3S	62.9E	51	29	83	01m45s
0535	1842 Jan 11	16:25:41	8	119	A	-0.8882	0.9151	75.8S	1.4E	27	288	710	06m15s
0536	1842 Jul 08	07:06:27	8	124	T	0.4727	1.0543	50.1N	83.6E	62	198	204	04m05s
0537	1842 Dec 31	19:04:24	8	129	A	-0.1727	0.9634	33.1S	103.2W	80	349	135	03m54s
0538	1843 Jun 27	19:17:03	8	134	H	-0.3037	1.0011	5.9N	111.0W	72	7	4	00m07s
0539	1843 Dec 21	05:03:26	8	139	T	0.5227	1.0165	8.0N	101.0E	58	175	66	01m43s
0540	1845 May 06	10:09:00	8	116	An	0.9945	0.9462	73.4N	110.5W	4	41	-	03m15s

Cat Num	Calendar Date	TD of Greatest Eclipse	ΔT s	Saros Num	Ecl Type	Gamma	Ecl Mag	Lat °	Long °	Sun Alt °	Sun Azm °	Path Width km	Central Line Dur
0541	1845 Oct 30	23:51:57	8	121	H	-0.8537	1.0005	69.1S	144.6E	31	39	3	00m02s
0542	1846 Apr 25	16:50:30	8	126	H	0.2038	1.0088	24.8N	76.2W	78	165	31	00m53s
0543	1846 Oct 20	07:46:11	8	131	A	-0.1506	0.9567	18.7S	57.3E	81	16	159	05m05s
0544	1847 Apr 15	06:16:13	8	136	T	-0.5339	1.0530	21.6S	95.1E	58	343	206	04m44s
0545	1847 Oct 09	09:00:22	8	141	A	0.5774	0.9290	27.7N	52.8E	55	199	323	08m35s
0546	1849 Feb 23	01:38:09	8	118	A	0.7474	0.9796	36.7N	144.3E	41	161	108	01m58s
0547	1849 Aug 18	05:40:49	8	123	T	-0.7343	1.0349	32.9S	83.5E	43	16	172	03m07s
0548	1850 Feb 12	06:29:37	9	128	A	0.0503	0.9345	11.0S	85.6E	87	166	245	08m35s
0549	1850 Aug 07	21:33:54	9	133	T	0.0215	1.0769	17.7N	141.8W	89	191	249	06m50s
0550	1851 Feb 01	05:54:27	9	138	A	-0.6413	0.9175	56.4S	106.9E	50	342	409	08m01s
0551	1851 Jul 28	14:33:42	9	143	T	0.7644	1.0577	68.0N	19.6W	40	201	296	03m41s
0552	1852 Dec 11	03:40:44	9	120	T	0.8551	1.0237	35.2N	133.9E	31	191	156	02m05s
0553	1853 Jun 06	20:07:21	9	125	A	-0.3685	0.9486	1.5N	117.9W	68	349	203	06m59s
0554	1853 Nov 30	19:15:39	9	130	T	0.1763	1.0485	12.0S	109.0W	80	194	164	04m28s
0555	1854 May 26	20:42:53	9	135	A	0.3918	0.9551	43.3N	140.0W	67	159	178	04m32s
0556	1854 Nov 20	09:56:58	9	140	H3	-0.5179	1.0144	48.9S	12.7E	59	27	57	01m07s
0557	1856 Apr 05	06:01:01	9	117	T	-0.7906	1.0539	39.1S	119.2E	38	323	285	03m56s
0558	1856 Sep 29	03:59:44	9	122	A	0.9420	0.9246	54.3N	169.1E	19	236	831	06m21s
0559	1857 Mar 25	22:29:38	9	127	T	-0.0892	1.0534	2.4S	153.4W	85	331	177	04m28s
0560	1857 Sep 18	05:36:05	9	132	A	0.1912	0.9659	11.6N	100.0E	79	210	125	03m34s
0561	1858 Mar 15	12:05:28	9	137	A	0.6461	0.9996	32.7N	20.9W	50	145	2	00m02s
0562	1858 Sep 07	14:09:29	9	142	T	-0.5609	1.0210	23.9S	49.8W	56	31	85	01m50s
0563	1860 Jan 23	00:27:31	9	119	A	-0.8969	0.9168	71.8S	117.2W	26	286	719	06m07s
0564	1860 Jul 18	14:26:24	9	124	T	0.5487	1.0500	52.5N	20.3W	56	205	198	03m39s
0565	1861 Jan 11	03:29:23	9	129	A	-0.1766	0.9664	31.8S	132.7E	80	344	123	03m30s
0566	1861 Jul 08	02:10:26	9	134	A	-0.2231	0.9979	10.0N	145.8E	77	12	7	00m14s
0567	1861 Dec 31	13:49:06	9	139	T	0.5187	1.0186	7.8N	31.6W	59	171	74	01m55s
0568	1863 Nov 11	08:09:03	9	121	A	-0.8759	0.9943	75.4S	15.2E	28	43	42	00m22s
0569	1864 May 06	00:16:48	9	126	H	0.2621	1.0146	31.6N	171.5E	75	168	52	01m25s
0570	1864 Oct 30	15:30:31	9	131	A	-0.1816	0.9514	24.3S	59.3W	79	14	181	05m41s
0571	1865 Apr 25	14:08:34	9	136	T	-0.4826	1.0584	14.8S	25.8W	61	346	219	05m23s
0572	1865 Oct 19	16:21:13	9	141	A	0.5366	0.9263	21.3N	60.2W	57	196	326	09m27s
0573	1867 Mar 06	09:46:47	9	118	A	0.7716	0.9787	42.3N	18.4E	39	157	118	01m57s
0574	1867 Aug 29	13:13:07	9	123	T	-0.7940	1.0344	41.1S	34.9W	37	21	189	02m51s
0575	1868 Feb 23	14:21:31	9	128	A	0.0706	0.9348	6.1S	33.0W	86	164	243	08m30s
0576	1868 Aug 18	05:12:10	9	133	Tm	-0.0443	1.0756	10.6N	102.3E	88	14	245	06m47s
0577	1869 Feb 11	13:46:39	9	138	A	-0.6251	0.9201	51.3S	9.7W	51	339	387	08m02s
0578	1869 Aug 07	22:01:05	9	143	T	0.6960	1.0551	59.1N	133.1W	46	202	254	03m48s
0579	1870 Dec 22	12:27:33	2	120	T	0.8585	1.0248	35.7N	1.5W	31	186	165	02m11s
0580	1871 Jun 18	02:35:01	2	125	A	-0.4550	0.9481	3.5S	144.7E	63	353	214	07m14s
0581	1871 Dec 12	04:03:38	2	130	T	0.1836	1.0465	12.7S	119.4E	80	190	157	04m23s
0582	1872 Jun 06	03:20:03	2	135	A	0.3095	0.9590	40.5N	124.8E	72	166	157	04m20s
0583	1872 Nov 30	18:29:33	2	140	H	-0.5081	1.0099	51.2S	111.8W	59	22	40	00m47s
0584	1874 Apr 16	14:00:53	2	117	T	-0.8364	1.0569	39.9S	0.9W	33	325	335	04m11s
0585	1874 Oct 10	11:13:33	2	122	An	0.9889	0.9193	58.6N	72.0E	7	244	-	06m28s
0586	1875 Apr 06	06:37:26	2	127	T	-0.1291	1.0547	0.2S	84.8E	83	332	182	04m37s
0587	1875 Sep 29	12:58:09	2	132	A	0.2427	0.9656	10.0N	10.1W	76	209	127	03m36s
0588	1876 Mar 25	20:05:06	2	137	A	0.6143	0.9999	34.8N	141.0W	52	144	1	00m01s
0589	1876 Sep 17	21:49:15	2	142	T	-0.5054	1.0220	24.6S	164.4W	60	32	86	01m53s
0590	1878 Feb 02	08:27:52	2	119	A	-0.9071	0.9191	67.9S	122.5E	24	286	729	05m59s
0591	1878 Jul 29	21:47:18	2	124	T	0.6232	1.0450	53.8N	124.0W	51	213	191	03m11s
0592	1879 Jan 22	11:53:08	2	129	A	-0.1824	0.9700	29.8S	8.5E	79	340	109	03m03s
0593	1879 Jul 19	09:04:32	2	134	Am	-0.1439	0.9942	13.0N	43.0E	82	16	20	00m39s
0594	1880 Jan 11	22:34:25	-3	139	T	0.5136	1.0212	8.3N	164.1W	59	166	84	02m07s
0595	1880 Jul 07	13:10:28	-3	144	A	-0.9406	0.9441	46.4S	33.4W	19	17	611	05m47s
0596	1881 Nov 21	16:31:10	-3	121	A	-0.8930	0.9887	81.2S	114.5W	26	46	90	00m43s
0597	1882 May 17	07:36:27	-3	126	T	0.3269	1.0200	38.4N	61.6E	71	171	72	01m50s
0598	1882 Nov 10	23:22:21	-3	131	A	-0.2056	0.9466	29.2S	177.0W	78	11	201	06m14s
0599	1883 May 06	21:53:49	-3	136	T	-0.4250	1.0634	8.1S	144.6W	65	349	229	05m58s
0600	1883 Oct 30	23:50:54	-3	141	A	0.5030	0.9238	15.6N	174.9W	60	193	331	10m17s

Cat Num	Calendar Date	TD of Greatest Eclipse	ΔT s	Saros Num	Ecl Type	Gamma	Ecl Mag	Lat °	Long °	Sun Alt °	Sun Azm °	Path Width km	Central Line Dur
0601	1885 Mar 16	17:45:42	-3	118	A	0.8030	0.9778	48.9N	106.1W	36	153	132	01m55s
0602	1885 Sep 08	20:51:52	-3	123	T	-0.8489	1.0332	49.6S	156.5W	32	27	211	02m31s
0603	1886 Mar 05	22:05:25	-3	128	A	0.0970	0.9357	0.5S	150.1W	84	163	240	08m20s
0604	1886 Aug 29	12:55:23	-3	133	T	-0.1059	1.0735	3.5N	15.3W	84	16	240	06m36s
0605	1887 Feb 22	21:33:03	-3	138	A	-0.6040	0.9232	45.7S	126.5W	53	338	362	08m01s
0606	1887 Aug 19	05:32:05	-3	143	T	0.6312	1.0518	50.6N	111.9E	51	202	221	03m50s
0607	1889 Jan 01	21:16:50	-3	120	T	0.8603	1.0262	36.7N	137.6W	30	181	175	02m17s
0608	1889 Jun 28	09:00:00	-3	125	A	-0.5431	0.9471	9.6S	47.3E	57	357	232	07m22s
0609	1889 Dec 22	12:54:14	-3	130	T	0.1888	1.0449	12.7S	12.8W	79	185	152	04m18s
0610	1890 Jun 17	09:55:05	-4	135	A	0.2247	0.9625	36.5N	29.3E	77	172	140	04m09s
0611	1890 Dec 12	03:05:28	-4	140	H	-0.5016	1.0059	52.8S	123.9E	60	15	24	00m28s
0612	1891 Jun 06	16:15:36	-4	145	A	0.9755	0.9981	74.5N	163.8E	12	45	33	00m06s
0613	1892 Apr 26	21:55:20	-4	117	T	-0.8870	1.0591	42.5S	119.4W	27	327	414	04m19s
0614	1893 Apr 16	14:36:11	-4	127	T	-0.1763	1.0556	1.3N	34.6W	80	334	186	04m47s
0615	1893 Oct 09	20:30:22	-4	132	A	0.2866	0.9652	8.1N	123.0W	73	208	130	03m41s
0616	1894 Apr 06	03:53:41	-4	137	H	0.5740	1.0001	36.7N	102.4E	55	144	1	00m01s
0617	1894 Sep 29	05:39:02	-4	142	T	-0.4574	1.0226	26.1S	78.5E	63	32	86	01m55s
0618	1896 Feb 13	16:23:13	-4	119	A	-0.9220	0.9218	64.6S	3.5E	22	287	761	05m48s
0619	1896 Aug 09	05:09:00	-4	124	T	0.6963	1.0392	54.4N	132.2E	46	220	182	02m43s
0620	1897 Feb 01	20:15:15	-4	129	A	-0.1903	0.9742	27.1S	115.7W	79	336	94	02m34s
0621	1897 Jul 29	15:56:58	-4	134	A	-0.0640	0.9899	15.3N	59.0W	86	20	35	01m05s
0622	1898 Jan 22	07:19:12	-4	139	T	0.5079	1.0244	9.5N	63.6E	59	162	96	02m21s
0623	1898 Jul 18	19:36:54	-4	144	A	-0.8547	0.9450	35.7S	130.1W	31	19	385	06m11s
0624	1899 Dec 03	00:57:28	-4	121	A	-0.9061	0.9836	86.6S	121.5E	25	43	140	01m01s
0625	1900 May 28	14:53:56	-2	126	T	0.3943	1.0249	44.8N	46.5W	67	175	92	02m10s
0626	1900 Nov 22	07:19:43	-2	131	A	-0.2245	0.9421	33.1S	64.8E	77	7	220	06m42s
0627	1901 May 18	05:33:48	-2	136	T	-0.3626	1.0680	1.7S	98.4E	69	353	238	06m29s
0628	1901 Nov 11	07:28:21	-2	141	A	0.4758	0.9216	10.8N	68.9E	62	190	336	11m01s
0629	1903 Mar 29	01:35:23	-2	118	A	0.8413	0.9767	56.2N	130.3E	32	147	153	01m53s
0630	1903 Sep 21	04:39:52	-2	123	T	-0.8967	1.0316	58.0S	77.2E	26	35	241	02m12s
0631	1904 Mar 17	05:40:44	-2	128	A	0.1299	0.9368	5.6N	94.7E	82	162	237	08m07s
0632	1904 Sep 09	20:44:21	-2	133	T	-0.1625	1.0709	3.7S	134.6W	81	17	234	06m20s
0633	1905 Mar 06	05:12:26	-2	138	A	-0.5768	0.9269	39.5S	117.4E	55	338	334	07m58s
0634	1905 Aug 30	13:07:26	-2	143	T	0.5708	1.0477	42.5N	4.3W	55	202	192	03m46s
0635	1907 Jan 14	06:05:43	-2	120	T	0.8628	1.0281	38.3N	86.4E	30	175	189	02m25s
0636	1907 Jul 10	15:24:32	-2	125	A	-0.6313	0.9456	16.9S	50.9W	51	2	258	07m23s
0637	1908 Jan 03	21:45:21	-2	130	T	0.1933	1.0437	11.8S	145.2W	79	180	149	04m14s
0638	1908 Jun 28	16:29:51	-2	135	A	0.1390	0.9655	31.4N	67.2W	82	177	126	04m00s
0639	1908 Dec 23	11:44:28	-2	140	H	-0.4985	1.0024	53.4S	0.6W	60	8	10	00m12s
0640	1909 Jun 17	23:18:38	-2	145	H	0.8957	1.0065	82.9N	123.5E	26	110	51	00m24s
0641	1910 May 09	05:42:13	11	117	T	-0.9437	1.0600	48.2S	125.2E	19	328	594	04m15s
0642	1911 Apr 28	22:27:22	11	127	T	-0.2294	1.0562	1.9N	151.9W	77	336	190	04m57s
0643	1911 Oct 22	04:13:02	11	132	A	0.3224	0.9650	6.3N	121.4E	71	206	133	03m47s
0644	1912 Apr 17	11:34:22	11	137	H	0.5280	1.0003	38.4N	11.3W	58	146	1	00m02s
0645	1912 Oct 10	13:36:14	11	142	T	-0.4149	1.0229	28.1S	40.1W	65	32	85	01m55s
0646	1914 Feb 25	00:13:01	11	119	A	-0.9416	0.9248	62.1S	113.3W	19	287	839	05m35s
0647	1914 Aug 21	12:34:27	11	124	T	0.7655	1.0328	54.5N	27.0E	40	227	170	02m14s
0648	1915 Feb 14	04:33:20	11	129	A	-0.2024	0.9789	24.0S	120.7E	78	333	77	02m04s
0649	1915 Aug 10	22:52:25	11	134	A	0.0124	0.9853	16.4N	161.4W	89	200	52	01m33s
0650	1916 Feb 03	16:00:21	11	139	T	0.4988	1.0280	11.1N	67.7W	60	158	108	02m36s
0651	1916 Jul 30	02:06:10	11	144	A	-0.7710	0.9447	29.0S	132.4E	39	22	313	06m24s
0652	1917 Dec 14	09:27:20	11	121	A	-0.9157	0.9791	88.0S	124.7E	23	271	189	01m17s
0653	1918 Jun 08	22:07:43	11	126	T	0.4658	1.0292	50.9N	152.1W	62	180	112	02m23s
0654	1918 Dec 03	15:22:01	11	131	A	-0.2387	0.9383	36.1S	53.7W	76	3	236	07m06s
0655	1919 May 29	13:08:55	11	136	T	-0.2955	1.0719	4.4N	16.7W	73	356	244	06m51s
0656	1919 Nov 22	15:14:12	11	141	A	0.4549	0.9198	6.9N	48.9W	63	186	341	11m37s
0657	1921 Apr 08	09:15:01	22	118	A	0.8869	0.9753	64.5N	5.6E	27	139	191	01m50s
0658	1921 Oct 01	12:35:58	22	123	T	-0.9383	1.0293	66.1S	56.1W	20	48	291	01m52s
0659	1922 Mar 28	13:05:26	22	128	A	0.1711	0.9381	12.3N	18.0W	80	162	233	07m50s
0660	1922 Sep 21	04:40:31	22	133	T	-0.2130	1.0678	10.7S	104.5E	78	18	226	05m59s

Cat Num	Calendar Date	TD of Greatest Eclipse	ΔT s	Saros Num	Ecl Type	Gamma	Ecl Mag	Lat °	Long °	Sun Alt °	Sun Azm °	Path Width km	Central Line Dur
0661	1923 Mar 17	12:44:58	22	138	A	-0.5438	0.9310	33.0S	2.3E	57	339	305	07m51s
0662	1923 Sep 10	20:47:29	22	143	T	0.5149	1.0430	34.7N	121.8W	59	201	167	03m37s
0663	1925 Jan 24	14:54:03	22	120	T	0.8661	1.0304	40.5N	49.6W	30	170	206	02m32s
0664	1925 Jul 20	21:48:42	22	125	A	-0.7193	0.9436	25.3S	150.0W	44	6	300	07m15s
0665	1926 Jan 14	06:36:58	22	130	T	0.1973	1.0430	10.1S	82.3E	79	176	147	04m11s
0666	1926 Jul 09	23:06:02	22	135	A	0.0538	0.9680	25.6N	165.1W	87	181	115	03m51s
0667	1927 Jan 03	20:22:53	22	140	A	-0.4956	0.9995	52.8S	124.8W	60	0	2	00m03s
0668	1927 Jun 29	06:23:27	22	145	T	0.8163	1.0128	78.1N	73.8E	35	167	77	00m50s
0669	1928 May 19	13:24:20	22	117	T-	-1.0048	1.0140	63.3S	22.5E	0	319	-	-
0670	1929 May 09	06:10:34	22	127	T	-0.2887	1.0562	1.6N	92.7E	73	339	193	05m07s
0671	1929 Nov 01	12:05:10	22	132	A	0.3514	0.9649	4.5N	3.1E	69	204	134	03m54s
0672	1930 Apr 28	19:03:34	24	137	H	0.4731	1.0003	39.4N	121.2W	62	149	1	00m01s
0673	1930 Oct 21	21:43:53	24	142	T	-0.3804	1.0230	30.5S	161.1W	67	31	84	01m55s
0674	1932 Mar 07	07:55:50	24	119	A	-0.9673	0.9277	60.7S	134.4E	14	285	1083	05m19s
0675	1932 Aug 31	20:03:41	24	124	T	0.8307	1.0257	54.5N	79.5W	34	232	155	01m45s
0676	1933 Feb 24	12:46:39	24	129	A	-0.2191	0.9841	20.8S	2.1W	77	331	58	01m32s
0677	1933 Aug 21	05:49:11	24	134	A	0.0869	0.9801	16.9N	95.9E	85	206	71	02m04s
0678	1934 Feb 14	00:38:41	24	139	T	0.4868	1.0321	13.2N	161.7E	61	155	123	02m53s
0679	1934 Aug 10	08:37:48	24	144	A	-0.6890	0.9436	24.5S	34.6E	46	25	280	06m33s
0680	1935 Dec 25	17:59:52	24	121	A	-0.9228	0.9753	83.5S	9.4E	22	258	234	01m30s
0681	1936 Jun 19	05:20:31	24	126	T	0.5389	1.0329	56.1N	104.7E	57	188	132	02m31s
0682	1936 Dec 13	23:28:12	24	131	A	-0.2493	0.9349	37.8S	172.6W	75	357	251	07m25s
0683	1937 Jun 08	20:41:02	24	136	T	-0.2253	1.0751	9.9N	130.5W	77	0	250	07m04s
0684	1937 Dec 02	23:05:45	24	141	A	0.4389	0.9184	4.0N	167.8W	64	182	344	12m00s
0685	1938 May 29	13:50:18	24	146	T	-0.9607	1.0552	52.7S	22.0W	16	354	675	04m05s
0686	1939 Apr 19	16:45:53	24	118	A	0.9388	0.9731	73.1N	129.1W	20	118	285	01m49s
0687	1939 Oct 12	20:40:23	24	123	T	-0.9737	1.0266	72.8S	155.1E	12	74	418	01m32s
0688	1940 Apr 07	20:21:21	24	128	A	0.2190	0.9394	19.2N	128.5W	77	163	230	07m30s
0689	1940 Oct 01	12:44:06	24	133	T	-0.2573	1.0645	17.5S	18.2W	75	18	218	05m35s
0690	1941 Mar 27	20:08:08	24	138	A	-0.5025	0.9355	26.2S	110.9W	60	341	276	07m41s
0691	1941 Sep 21	04:34:03	24	143	T	0.4649	1.0379	27.3N	119.1E	62	200	143	03m22s
0692	1943 Feb 04	23:38:10	24	120	T	0.8734	1.0331	43.6N	175.1E	29	165	229	02m39s
0693	1943 Aug 01	04:16:13	24	125	A	-0.8041	0.9409	34.8S	108.6E	36	11	367	06m59s
0694	1944 Jan 25	15:26:42	24	130	T	0.2025	1.0428	7.6S	50.2W	78	172	146	04m09s
0695	1944 Jul 20	05:43:13	24	135	A	-0.0313	0.9700	19.0N	95.7E	88	6	108	03m42s
0696	1945 Jan 14	05:01:43	24	140	A	-0.4937	0.9970	51.1S	110.2E	60	354	12	00m15s
0697	1945 Jul 09	13:27:45	24	145	T	0.7356	1.0180	70.0N	17.2W	42	184	92	01m15s
0698	1947 May 20	13:47:47	24	127	T	-0.3528	1.0557	0.2N	21.4W	69	343	196	05m13s
0699	1947 Nov 12	20:05:37	24	132	A	0.3743	0.9650	3.0N	117.4W	68	200	135	03m59s
0700	1948 May 09	02:26:04	24	137	A	0.4133	0.9999	39.8N	131.2E	65	153	0	00m00s
0701	1948 Nov 01	05:59:18	24	142	T	-0.3517	1.0231	33.1S	76.2E	69	28	84	01m56s
0702	1950 Mar 18	15:32:01	29	119	A-	-0.9988	0.9620	60.9S	40.9E	0	268	-	-
0703	1950 Sep 12	03:38:47	29	124	T	0.8903	1.0182	54.8N	172.3E	27	236	134	01m14s
0704	1951 Mar 07	20:53:40	29	129	A	-0.2420	0.9896	17.7S	123.5W	76	330	38	00m59s
0705	1951 Sep 01	12:51:51	29	134	A	0.1557	0.9747	16.5N	8.5W	81	208	91	02m36s
0706	1952 Feb 25	09:11:35	29	139	T	0.4697	1.0366	15.6N	32.7E	62	152	138	03m09s
0707	1952 Aug 20	15:13:35	29	144	A	-0.6102	0.9420	21.7S	64.1W	52	27	264	06m40s
0708	1954 Jan 05	02:32:01	29	121	A	-0.9296	0.9720	79.1S	120.8W	21	260	278	01m42s
0709	1954 Jun 30	12:32:38	29	126	T	0.6134	1.0357	60.5N	4.2E	52	197	153	02m35s
0710	1954 Dec 25	07:36:42	29	131	A	-0.2576	0.9323	38.4S	68.2E	75	352	262	07m39s
0711	1955 Jun 20	04:10:42	29	136	T	-0.1528	1.0776	14.8N	117.0E	81	5	254	07m08s
0712	1955 Dec 14	07:02:25	29	141	A	0.4266	0.9176	2.1N	72.2E	65	178	346	12m09s
0713	1956 Jun 08	21:20:39	29	146	T	-0.8934	1.0581	40.8S	140.8W	26	0	429	04m45s
0714	1957 Apr 30	00:05:28	29	118	A+	0.9992	0.9799	70.6N	40.3E	0	41	-	-
0715	1957 Oct 23	04:54:02	29	123	T-	-1.0022	1.0013	71.2S	23.1W	0	127	-	-
0716	1958 Apr 19	03:27:17	29	128	A	0.2750	0.9408	26.5N	123.5E	74	164	228	07m07s
0717	1958 Oct 12	20:55:28	29	133	T	-0.2951	1.0608	24.0S	142.4W	73	18	209	05m11s
0718	1959 Apr 08	03:24:08	29	138	A	-0.4546	0.9401	19.1S	137.6E	63	343	247	07m26s
0719	1959 Oct 02	12:27:00	29	143	T	0.4207	1.0325	20.4N	1.5W	65	199	120	03m02s
0720	1961 Feb 15	08:19:48	33	120	T	0.8830	1.0360	47.4N	40.0E	28	159	258	02m45s

Cat Num	Calendar Date	TD of Greatest Eclipse	ΔT s	Saros Num	Ecl Type	Gamma	Ecl Mag	Lat °	Long °	Sun Alt °	Sun Azm °	Path Width km	Central Line Dur
0721	1961 Aug 11	10:46:47	33	125	A	-0.8859	0.9375	45.8S	4.0E	27	17	499	06m35s
0722	1962 Feb 05	00:12:38	33	130	T	0.2107	1.0430	4.2S	178.1E	78	169	147	04m08s
0723	1962 Jul 31	12:25:33	33	135	Am	-0.1130	0.9716	12.0N	5.7W	84	9	103	03m33s
0724	1963 Jan 25	13:37:12	33	140	A	-0.4898	0.9951	48.2S	15.0W	60	348	20	00m25s
0725	1963 Jul 20	20:36:13	33	145	T	0.6571	1.0224	61.7N	119.6W	49	191	101	01m40s
0726	1965 May 30	21:17:31	33	127	T	-0.4225	1.0544	2.5S	133.8W	65	347	198	05m15s
0727	1965 Nov 23	04:14:51	33	132	A	0.3906	0.9656	1.7N	119.8E	67	197	134	04m02s
0728	1966 May 20	09:39:02	33	137	A	0.3467	0.9992	39.2N	26.4E	70	158	3	00m05s
0729	1966 Nov 12	14:23:28	33	142	T	-0.3300	1.0234	35.6S	48.2W	71	25	84	01m57s
0730	1967 Nov 02	05:38:56	33	152	T-	-1.0007	1.0126	62.0S	27.9W	0	122	-	-
0731	1968 Sep 22	11:18:46	33	124	T	0.9451	1.0099	56.2N	64.0E	19	240	104	00m40s
0732	1969 Mar 18	04:54:57	33	129	A	-0.2704	0.9954	14.8S	116.3E	74	330	16	00m26s
0733	1969 Sep 11	19:58:59	33	134	A	0.2201	0.9690	15.6N	114.2W	77	210	114	03m11s
0734	1970 Mar 07	17:38:30	40	139	T	0.4473	1.0414	18.2N	94.7W	63	150	153	03m28s
0735	1970 Aug 31	21:55:30	40	144	A	-0.5364	0.9400	20.3S	164.0W	57	29	258	06m47s
0736	1972 Jan 16	11:03:22	40	121	A	-0.9365	0.9692	74.9S	107.7E	20	263	321	01m53s
0737	1972 Jul 10	19:46:38	40	126	T	0.6872	1.0379	63.5N	94.2W	46	209	175	02m36s
0738	1973 Jan 04	15:46:21	40	131	A	-0.2644	0.9303	37.9S	51.2W	74	346	271	07m49s
0739	1973 Jun 30	11:38:41	40	136	T	-0.0785	1.0792	18.8N	5.6E	86	9	256	07m04s
0740	1973 Dec 24	15:02:44	40	141	A	0.4171	0.9174	1.1N	48.5W	65	174	345	12m02s
0741	1974 Jun 20	04:48:04	40	146	T	-0.8239	1.0592	32.1S	103.7E	34	5	344	05m09s
0742	1976 Apr 29	10:24:18	45	128	A	0.3378	0.9421	34.0N	18.3E	70	165	227	06m41s
0743	1976 Oct 23	05:13:45	46	133	T	-0.3270	1.0572	30.0S	92.3E	71	17	199	04m46s
0744	1977 Apr 18	10:31:30	46	138	A	-0.3990	0.9449	11.9S	28.3E	66	345	220	07m04s
0745	1977 Oct 12	20:27:27	46	143	T	0.3836	1.0269	14.1N	123.6W	67	197	99	02m37s
0746	1979 Feb 26	16:55:06	46	120	T	0.8981	1.0391	52.1N	94.5W	26	153	298	02m49s
0747	1979 Aug 22	17:22:38	46	125	A	-0.9632	0.9329	59.6S	108.6W	15	29	953	06m03s
0748	1980 Feb 16	08:54:01	51	130	T	0.2224	1.0434	0.1S	47.1E	77	166	149	04m08s
0749	1980 Aug 10	19:12:21	51	135	A	-0.1915	0.9727	4.6N	108.9W	79	12	100	03m23s
0750	1981 Feb 04	22:09:24	51	140	A	-0.4838	0.9938	44.4S	140.8W	61	344	25	00m33s
0751	1981 Jul 31	03:46:37	51	145	T	0.5792	1.0258	53.3N	134.1E	54	195	108	02m02s
0752	1983 Jun 11	04:43:33	51	127	T	-0.4947	1.0524	6.2S	114.2E	60	351	199	05m11s
0753	1983 Dec 04	12:31:15	51	132	A	0.4015	0.9666	0.9N	4.7W	66	192	131	04m01s
0754	1984 May 30	16:45:42	51	137	A	0.2755	0.9980	37.5N	76.8W	74	163	7	00m11s
0755	1984 Nov 22	22:54:17	51	142	T	-0.3132	1.0237	37.8S	173.7W	72	21	85	02m00s
0756	1985 Nov 12	14:11:27	54	152	T	-0.9795	1.0388	68.6S	142.6W	11	111	690	01m59s
0757	1986 Oct 03	19:06:15	54	124	H	0.9931	1.0000	59.9N	37.1W	5	252	1	00m00s
0758	1987 Mar 29	12:49:47	54	129	H	-0.3053	1.0013	12.3S	2.3W	72	331	5	00m08s
0759	1987 Sep 23	03:12:22	54	134	A	0.2787	0.9634	14.3N	138.4E	74	210	137	03m49s
0760	1988 Mar 18	01:58:56	54	139	T	0.4188	1.0464	20.7N	140.0E	65	149	169	03m46s
0761	1988 Sep 11	04:44:29	54	144	A	-0.4681	0.9377	20.0S	94.4E	62	31	258	06m57s
0762	1990 Jan 26	19:31:24	57	121	A	-0.9457	0.9670	71.0S	22.2W	18	266	373	02m03s
0763	1990 Jul 22	03:03:07	57	126	T	0.7597	1.0391	65.2N	168.9E	40	222	201	02m33s
0764	1991 Jan 15	23:53:51	57	131	A	-0.2727	0.9290	36.4S	170.4W	74	341	277	07m53s
0765	1991 Jul 11	19:07:01	57	136	Tm	-0.0041	1.0800	22.0N	105.2W	90	30	258	06m53s
0766	1992 Jan 04	23:05:37	57	141	A	0.4091	0.9179	1.0N	169.7W	66	169	340	11m41s
0767	1992 Jun 30	12:11:22	57	146	T	-0.7512	1.0592	25.2S	9.5W	41	10	294	05m21s
0768	1994 May 10	17:12:26	57	128	A	0.4077	0.9431	41.5N	84.1W	66	168	230	06m13s
0769	1994 Nov 03	13:40:06	57	133	T	-0.3522	1.0535	35.4S	34.2W	69	15	189	04m23s
0770	1995 Apr 29	17:33:21	61	138	A	-0.3382	0.9497	4.8S	79.4W	70	348	195	06m37s
0771	1995 Oct 24	04:33:31	61	143	T	0.3518	1.0213	8.4N	113.2E	69	195	78	02m10s
0772	1997 Mar 09	01:24:51	61	120	T	0.9183	1.0420	57.8N	130.7E	23	146	356	02m50s
0773	1998 Feb 26	17:29:27	61	130	T	0.2391	1.0441	4.7N	82.7W	76	164	151	04m09s
0774	1998 Aug 22	02:07:11	61	135	A	-0.2644	0.9734	3.0S	145.4E	75	14	99	03m14s
0775	1999 Feb 16	06:34:38	61	140	A	-0.4726	0.9928	39.8S	93.9E	62	342	29	00m40s
0776	1999 Aug 11	11:04:09	61	145	T	0.5062	1.0286	45.1N	24.3E	59	197	112	02m23s
0777	2001 Jun 21	12:04:46	64	127	T	-0.5701	1.0495	11.3S	2.7E	55	355	200	04m57s
0778	2001 Dec 14	20:53:01	64	132	A	0.4089	0.9681	0.6N	130.7W	66	188	126	03m53s
0779	2002 Jun 10	23:45:22	64	137	A	0.1993	0.9962	34.5N	178.6W	78	169	13	00m23s
0780	2002 Dec 04	07:32:16	64	142	T	-0.3020	1.0244	39.5S	59.6E	72	16	87	02m04s

Cat Num	Calendar Date	TD of Greatest Eclipse	ΔT s	Saros Num	Ecl Type	Gamma	Ecl Mag	Lat °	Long °	Sun Alt °	Sun Azm °	Path Width km	Central Line Dur
0781	2003 May 31	04:09:23	64	147	An	0.9960	0.9384	66.6N	24.5W	3	35	-	03m37s
0782	2003 Nov 23	22:50:22	64	152	T	-0.9638	1.0379	72.7S	88.4E	15	111	496	01m57s
0783	2005 Apr 08	20:36:51	65	129	H	-0.3473	1.0074	10.6S	119.0W	70	332	27	00m42s
0784	2005 Oct 03	10:32:47	65	134	A	0.3306	0.9576	12.9N	28.7E	71	209	162	04m32s
0785	2006 Mar 29	10:12:23	65	139	T	0.3843	1.0515	23.2N	16.7E	67	149	184	04m07s
0786	2006 Sep 22	11:41:16	65	144	A	-0.4062	0.9352	20.6S	9.1W	66	31	261	07m09s
0787	2008 Feb 07	03:56:10	65	121	A	-0.9570	0.9650	67.6S	150.5W	16	269	444	02m12s
0788	2008 Aug 01	10:22:12	65	126	T	0.8307	1.0394	65.7N	72.3E	34	235	237	02m27s
0789	2009 Jan 26	07:59:45	66	131	A	-0.2820	0.9282	34.1S	70.2E	73	337	280	07m54s
0790	2009 Jul 22	02:36:25	66	136	T	0.0698	1.0799	24.2N	144.1E	86	198	258	06m39s
0791	2010 Jan 15	07:07:39	66	141	A	0.4002	0.9190	1.6N	69.3E	66	165	333	11m08s
0792	2010 Jul 11	19:34:38	66	146	T	-0.6788	1.0580	19.7S	121.9W	47	14	259	05m20s
0793	2012 May 20	23:53:54	67	128	A	0.4828	0.9439	49.1N	176.3E	61	171	237	05m46s
0794	2012 Nov 13	22:12:55	67	133	T	-0.3719	1.0500	40.0S	161.3W	68	11	179	04m02s
0795	2013 May 10	00:26:20	67	138	A	-0.2694	0.9544	2.2N	175.5E	74	350	173	06m03s
0796	2013 Nov 03	12:47:36	67	143	H3	0.3271	1.0159	3.5N	11.7W	71	192	58	01m40s
0797	2014 Apr 29	06:04:33	67	148	A-	-1.0000	0.9868	70.6S	131.3E	0	319	-	-
0798	2015 Mar 20	09:46:47	68	120	T	0.9454	1.0446	64.4N	6.6W	18	135	463	02m47s
0799	2016 Mar 09	01:58:19	68	130	T	0.2609	1.0450	10.1N	148.8E	75	162	155	04m09s
0800	2016 Sep 01	09:08:02	68	135	A	-0.3330	0.9736	10.7S	37.8E	70	16	100	03m06s
0801	2017 Feb 26	14:54:33	69	140	A	-0.4578	0.9922	34.7S	31.2W	63	340	31	00m44s
0802	2017 Aug 21	18:26:40	69	145	T	0.4367	1.0306	37.0N	87.7W	64	198	115	02m40s
0803	2019 Jul 02	19:24:07	69	127	T	-0.6466	1.0459	17.4S	109.0W	50	359	201	04m33s
0804	2019 Dec 26	05:18:53	69	132	A	0.4135	0.9701	1.0N	102.2E	66	184	118	03m39s
0805	2020 Jun 21	06:41:15	69	137	Am	0.1209	0.9940	30.5N	79.7E	83	174	21	00m38s
0806	2020 Dec 14	16:14:39	69	142	T	-0.2939	1.0254	40.3S	68.0W	73	10	90	02m10s
0807	2021 Jun 10	10:43:07	69	147	A	0.9152	0.9435	80.8N	66.8W	23	90	527	03m51s
0808	2021 Dec 04	07:34:38	69	152	T	-0.9526	1.0367	76.8S	46.2W	17	115	419	01m54s
0809	2023 Apr 20	04:17:56	69	129	H	-0.3952	1.0132	9.6S	125.8E	67	334	49	01m16s
0810	2023 Oct 14	18:00:41	69	134	A	0.3753	0.9520	11.4N	83.1W	68	208	187	05m17s
0811	2024 Apr 08	18:18:29	69	139	T	0.3431	1.0566	25.3N	104.1W	70	149	198	04m28s
0812	2024 Oct 02	18:46:13	69	144	A	-0.3509	0.9326	22.0S	114.5W	69	31	266	07m25s
0813	2026 Feb 17	12:13:06	69	121	A	-0.9743	0.9630	64.7S	86.7E	12	268	616	02m20s
0814	2026 Aug 12	17:47:06	69	126	T	0.8977	1.0386	65.2N	25.2W	26	248	294	02m18s
0815	2027 Feb 06	16:00:48	69	131	A	-0.2952	0.9281	31.3S	48.5W	73	334	282	07m51s
0816	2027 Aug 02	10:07:50	69	136	T	0.1421	1.0790	25.5N	33.2E	82	202	258	06m23s
0817	2028 Jan 26	15:08:59	69	141	A	0.3901	0.9208	3.0N	51.6W	67	161	323	10m27s
0818	2028 Jul 22	02:56:40	69	146	T	-0.6056	1.0560	15.6S	126.7E	53	17	230	05m10s
0819	2030 Jun 01	06:29:13	70	128	A	0.5626	0.9443	56.5N	80.0E	55	176	250	05m21s
0820	2030 Nov 25	06:51:37	70	133	T	-0.3867	1.0468	43.6S	71.2E	67	7	169	03m44s
0821	2031 May 21	07:16:04	70	138	A	-0.1970	0.9589	8.9N	71.7E	79	354	152	05m26s
0822	2031 Nov 14	21:07:31	70	143	H	0.3078	1.0106	0.6S	137.7W	72	189	38	01m08s
0823	2032 May 09	13:26:42	71	148	A	-0.9375	0.9957	51.3S	7.1W	20	345	44	00m22s
0824	2033 Mar 30	18:02:36	71	120	T	0.9778	1.0462	71.3N	155.9W	11	111	781	02m37s
0825	2034 Mar 20	10:18:45	71	130	T	0.2894	1.0458	16.1N	22.2E	73	162	159	04m09s
0826	2034 Sep 12	16:19:28	71	135	A	-0.3936	0.9736	18.2S	72.6W	67	18	102	02m58s
0827	2035 Mar 09	23:05:54	72	140	A	-0.4368	0.9919	29.0S	155.0W	64	340	31	00m48s
0828	2035 Sep 02	01:56:46	72	145	T	0.3727	1.0320	29.1N	158.0E	68	199	116	02m54s
0829	2037 Jul 13	02:40:36	72	127	T	-0.7246	1.0413	24.8S	139.0E	43	3	201	03m58s
0830	2038 Jan 05	13:47:11	72	132	A	0.4169	0.9728	2.1N	25.5W	65	179	107	03m18s
0831	2038 Jul 02	13:32:55	72	137	A	0.0398	0.9911	25.4N	21.9W	88	179	31	01m00s
0832	2038 Dec 26	01:00:10	72	142	T	-0.2881	1.0269	40.3S	163.9E	73	5	95	02m18s
0833	2039 Jun 21	17:12:54	73	147	A	0.8312	0.9454	78.9N	102.2W	33	153	365	04m05s
0834	2039 Dec 15	16:23:46	73	152	T	-0.9458	1.0356	80.9S	172.7E	18	123	380	01m51s
0835	2041 Apr 30	11:52:21	73	129	T	-0.4492	1.0189	9.6S	12.1E	63	337	72	01m51s
0836	2041 Oct 25	01:36:22	73	134	A	0.4133	0.9467	9.9N	162.8E	66	206	213	06m07s
0837	2042 Apr 20	02:17:30	73	139	T	0.2956	1.0614	27.0N	137.3E	73	151	210	04m51s
0838	2042 Oct 14	02:00:42	73	144	A	-0.3030	0.9301	23.7S	137.8E	72	30	273	07m44s
0839	2043 Apr 09	18:57:49	74	149	T+	1.0031	1.0096	61.3N	151.9E	0	74	-	-
0840	2043 Oct 03	03:01:49	74	154	A-	-1.0102	0.9497	61.0S	35.2E	0	98	-	-

Cat Num	Calendar Date	TD of Greatest Eclipse	ΔT s	Saros Num	Ecl Type	Gamma	Ecl Mag	Lat °	Long °	Sun Alt °	Sun Azm °	Path Width km	Central Line Dur
0841	2044 Feb 28	20:24:39	74	121	As	-0.9954	0.9600	62.2S	25.7W	4	260	-	02m27s
0842	2044 Aug 23	01:17:02	74	126	T	0.9613	1.0364	64.3N	120.5W	15	264	453	02m04s
0843	2045 Feb 16	23:56:07	74	131	A	-0.3125	0.9285	28.3S	166.3W	72	331	281	07m47s
0844	2045 Aug 12	17:42:39	74	136	T	0.2116	1.0774	25.9N	78.6W	78	206	256	06m06s
0845	2046 Feb 05	23:06:26	74	141	A	0.3765	0.9232	4.8N	171.5W	68	157	310	09m42s
0846	2046 Aug 02	10:21:13	74	146	T	-0.5350	1.0531	12.7S	15.1E	58	21	206	04m51s
0847	2048 Jun 11	12:58:53	75	128	A	0.6468	0.9441	63.7N	11.6W	49	184	271	04m58s
0848	2048 Dec 05	15:35:27	75	133	T	-0.3973	1.0440	46.1S	56.4W	66	1	160	03m28s
0849	2049 May 31	13:59:59	75	138	A	-0.1187	0.9631	15.3N	29.9W	83	358	134	04m45s
0850	2049 Nov 25	05:33:48	75	143	H	0.2943	1.0057	3.8S	95.2E	73	185	21	00m38s
0851	2050 May 20	20:42:50	76	148	H	-0.8688	1.0038	40.1S	123.8W	29	352	27	00m21s
0852	2052 Mar 30	18:31:53	76	130	T	0.3239	1.0466	22.4N	102.6W	71	161	164	04m08s
0853	2052 Sep 22	23:39:10	76	135	A	-0.4480	0.9734	25.7S	174.9E	63	20	106	02m51s
0854	2053 Mar 20	07:08:19	76	140	A	-0.4089	0.9919	23.0S	82.9E	66	341	31	00m50s
0855	2053 Sep 12	09:34:09	76	145	T	0.3140	1.0328	21.5N	41.6E	72	199	116	03m04s
0856	2055 Jul 24	09:57:50	77	127	T	-0.8012	1.0359	33.3S	25.7E	37	8	202	03m17s
0857	2056 Jan 16	22:16:45	77	132	A	0.4199	0.9760	3.9N	153.6W	65	175	95	02m52s
0858	2056 Jul 12	20:21:59	77	137	A	-0.0426	0.9878	19.4N	123.9W	88	3	43	01m26s
0859	2057 Jan 05	09:47:52	77	142	T	-0.2837	1.0287	39.2S	35.1E	73	359	102	02m29s
0860	2057 Jul 01	23:40:15	78	147	A	0.7455	0.9464	71.5N	176.3W	41	177	298	04m22s
0861	2057 Dec 26	01:14:35	78	152	T	-0.9405	1.0348	84.9S	21.7E	19	141	355	01m50s
0862	2059 May 11	19:22:16	78	129	T	-0.5080	1.0242	10.7S	100.5W	59	340	95	02m23s
0863	2059 Nov 05	09:18:15	78	134	A	0.4454	0.9417	8.7N	46.9E	63	203	238	07m00s
0864	2060 Apr 30	10:10:00	78	139	T	0.2422	1.0660	28.0N	20.8E	76	154	222	05m15s
0865	2060 Oct 24	09:24:10	79	144	A	-0.2625	0.9277	25.8S	28.0E	75	28	281	08m06s
0866	2061 Apr 20	02:56:49	79	149	T	0.9578	1.0476	64.5N	59.0E	16	97	559	02m37s
0867	2061 Oct 13	10:32:10	79	154	A	-0.9639	0.9469	62.1S	54.5W	15	79	743	03m41s
0868	2063 Feb 28	07:43:30	79	131	A	-0.3360	0.9293	25.2S	77.6E	70	329	280	07m41s
0869	2063 Aug 24	01:22:11	80	136	T	0.2771	1.0750	25.6N	168.3E	74	209	252	05m49s
0870	2064 Feb 17	07:00:23	80	141	A	0.3596	0.9262	7.0N	69.6E	69	154	295	08m56s
0871	2064 Aug 12	17:46:06	80	146	T	-0.4652	1.0495	10.9S	96.2W	62	24	184	04m28s
0872	2066 Jun 22	19:25:48	80	128	A	0.7330	0.9435	70.1N	96.6W	43	198	309	04m40s
0873	2066 Dec 17	00:23:40	81	133	T	-0.4043	1.0416	47.4S	175.6E	66	355	152	03m14s
0874	2067 Jun 11	20:42:26	81	138	A	-0.0387	0.9670	21.0N	130.4W	88	2	119	04m05s
0875	2067 Dec 06	14:03:43	81	143	H	0.2845	1.0011	6.0S	32.6W	74	181	4	00m08s
0876	2068 May 31	03:56:39	81	148	T	-0.7970	1.0110	31.0S	123.0E	37	357	63	01m06s
0877	2070 Apr 11	02:36:09	82	130	T	0.3652	1.0472	29.1N	134.8E	68	162	168	04m04s
0878	2070 Oct 04	07:08:57	82	135	A	-0.4950	0.9731	32.8S	60.2E	60	21	110	02m44s
0879	2071 Mar 31	15:01:06	82	140	A	-0.3739	0.9919	16.7S	37.2W	68	342	31	00m52s
0880	2071 Sep 23	17:20:28	82	145	T	0.2620	1.0333	14.2N	77.0W	75	198	116	03m11s
0881	2072 Sep 12	08:59:20	83	155	T	0.9655	1.0558	69.8N	101.8E	14	240	732	03m13s
0882	2073 Aug 03	17:15:23	83	127	T	-0.8763	1.0294	43.2S	89.6W	28	14	206	02m29s
0883	2074 Jan 27	06:44:15	83	132	A	0.4251	0.9798	6.6N	78.5E	65	171	79	02m21s
0884	2074 Jul 24	03:10:32	83	137	A	-0.1242	0.9838	12.8N	133.5E	83	7	58	01m57s
0885	2075 Jan 16	18:36:04	83	142	T	-0.2799	1.0311	37.2S	94.4W	74	354	110	02m42s
0886	2075 Jul 13	06:05:44	84	147	A	0.6583	0.9467	63.1N	94.9E	49	186	262	04m45s
0887	2076 Jan 06	10:07:27	84	152	T	-0.9373	1.0342	87.2S	174.0W	20	203	340	01m49s
0888	2077 May 22	02:46:05	84	129	T	-0.5725	1.0290	13.1S	148.0E	55	343	119	02m54s
0889	2077 Nov 15	17:07:56	84	134	A	0.4705	0.9371	7.8N	71.1W	62	199	262	07m54s
0890	2078 May 11	17:56:55	85	139	T	0.1838	1.0701	28.1N	94.0W	79	158	232	05m40s
0891	2078 Nov 04	16:55:44	85	144	A	-0.2285	0.9255	27.8S	83.6W	77	25	287	08m29s
0892	2079 May 01	10:50:13	85	149	T	0.9081	1.0512	66.2N	46.6W	24	108	406	02m55s
0893	2079 Oct 24	18:11:21	85	154	A	-0.9243	0.9484	63.4S	160.9W	22	72	495	03m39s
0894	2081 Mar 10	15:23:31	86	131	A	-0.3653	0.9304	22.4S	37.0W	68	329	277	07m36s
0895	2081 Sep 03	09:07:31	86	136	T	0.3379	1.0720	24.6N	53.3E	70	211	247	05m33s
0896	2082 Feb 27	14:47:00	86	141	A	0.3361	0.9298	9.4N	47.4W	70	152	277	08m12s
0897	2082 Aug 24	01:16:21	86	146	T	-0.4004	1.0452	10.3S	151.5E	66	26	163	04m01s
0898	2084 Jul 03	01:50:26	87	128	A	0.8208	0.9421	75.0N	169.4W	35	222	377	04m25s
0899	2084 Dec 27	09:13:48	87	133	T	-0.4094	1.0396	47.3S	47.4E	66	349	146	03m04s
0900	2085 Jun 22	03:21:16	87	138	A	0.0453	0.9704	26.2N	130.9E	87	186	106	03m29s

Cat Num	Calendar Date	TD of Greatest Eclipse	ΔT s	Saros Num	Ecl Type	Gamma	Ecl Mag	Lat °	Long °	Sun Alt °	Sun Azm °	Path Width km	Central Line Dur
0901	2085 Dec 16	22:37:48	88	143	A	0.2786	0.9971	7.3S	161.1W	74	176	10	00m19s
0902	2086 Jun 11	11:07:14	88	148	T	-0.7215	1.0174	23.2S	12.1E	44	2	86	01m48s
0903	2088 Apr 21	10:31:49	89	130	T	0.4135	1.0474	36.0N	14.7E	65	163	173	03m58s
0904	2088 Oct 14	14:48:05	89	135	A	-0.5349	0.9727	39.7S	56.4W	57	21	115	02m38s
0905	2089 Apr 10	22:44:41	89	140	A	-0.3319	0.9919	10.2S	155.1W	71	344	30	00m53s
0906	2089 Oct 04	01:15:23	89	145	T	0.2167	1.0333	7.4N	162.4E	77	197	115	03m14s
0907	2090 Sep 23	16:56:36	90	155	T	0.9157	1.0562	60.7N	40.9W	23	218	463	03m36s
0908	2091 Aug 15	00:34:43	90	127	T	-0.9490	1.0216	55.6S	150.1E	18	23	237	01m38s
0909	2092 Feb 07	15:10:20	90	132	A	0.4322	0.9840	9.9N	49.1W	64	168	62	01m48s
0910	2092 Aug 03	09:59:33	90	137	A	-0.2044	0.9794	5.6N	29.9E	78	10	75	02m31s
0911	2093 Jan 27	03:22:16	91	142	T	-0.2737	1.0340	34.1S	136.0E	74	350	119	02m58s
0912	2093 Jul 23	12:32:04	91	147	A	0.5717	0.9463	54.6N	0.9E	55	191	241	05m11s
0913	2094 Jan 16	18:59:03	91	152	T	-0.9333	1.0342	84.8S	11.0W	21	267	329	01m52s
0914	2095 Jun 02	10:07:40	92	129	T	-0.6396	1.0332	16.7S	36.7E	50	347	145	03m18s
0915	2095 Nov 27	01:02:57	92	134	A	0.4903	0.9330	7.2N	169.4E	61	195	285	08m47s
0916	2096 May 22	01:37:14	92	139	T	0.1196	1.0737	27.3N	153.0E	83	162	241	06m06s
0917	2096 Nov 15	00:36:15	92	144	A	-0.2018	0.9237	29.7S	162.9E	78	22	294	08m53s
0918	2097 May 11	18:34:31	92	149	T	0.8516	1.0538	67.4N	150.0W	31	121	339	03m10s
0919	2097 Nov 04	02:01:25	93	154	A	-0.8926	0.9494	65.8S	86.3E	26	68	411	03m36s
0920	2099 Mar 21	22:54:32	93	131	A	-0.4016	0.9318	20.0S	149.5W	66	329	275	07m32s
0921	2099 Sep 14	16:57:53	93	136	T	0.3942	1.0684	23.4N	63.3W	67	211	241	05m18s
0922	2100 Mar 10	22:28:11	94	141	A	0.3077	0.9338	12.0N	162.9W	72	151	257	07m29s
0923	2100 Sep 04	08:49:20	94	146	T	-0.3384	1.0402	10.5S	38.5E	70	28	142	03m32s
0924	2101 Feb 28	02:16:26	94	151	An	0.9964	0.9609	60.5N	79.6E	3	111	-	02m44s
0925	2102 Jul 15	08:15:14	95	128	A	0.9080	0.9398	75.9N	133.7E	24	261	539	04m14s
0926	2103 Jan 08	18:04:21	95	133	T	-0.4140	1.0381	46.1S	81.3W	65	342	140	02m57s
0927	2103 Jul 04	10:01:48	95	138	Am	0.1285	0.9734	30.3N	32.7E	82	191	96	02m57s
0928	2103 Dec 29	07:13:18	95	143	A	0.2747	0.9936	7.5S	70.0E	74	172	23	00m43s
0929	2104 Jun 22	18:16:21	96	148	T	-0.6437	1.0231	16.6S	97.3W	50	6	103	02m26s
0930	2104 Dec 17	13:48:27	96	153	A+	1.0120	0.9381	66.4N	37.1W	0	171	-	-
0931	2106 May 03	18:19:19	97	130	T	0.4681	1.0472	43.1N	102.8W	62	164	178	03m47s
0932	2106 Oct 26	22:37:40	97	135	A	-0.5671	0.9725	45.9S	174.6W	55	20	119	02m32s
0933	2107 Apr 23	06:18:41	97	140	A	-0.2828	0.9918	3.6S	89.4E	74	346	30	00m56s
0934	2107 Oct 16	09:18:27	97	145	T	0.1778	1.0332	1.1N	40.0E	80	196	114	03m16s
0935	2108 Oct 05	01:01:20	98	155	T	0.8722	1.0551	52.5N	172.5W	29	209	371	03m50s
0936	2110 Feb 18	23:31:35	98	132	A	0.4438	0.9888	14.1N	175.8W	64	165	44	01m12s
0937	2110 Aug 15	16:50:45	99	137	A	-0.2819	0.9746	2.0S	74.8W	74	13	94	03m07s
0938	2111 Feb 08	12:05:33	99	142	T	-0.2650	1.0374	30.2S	6.3E	74	346	130	03m17s
0939	2111 Aug 04	19:00:22	99	147	A	0.4867	0.9455	46.0N	95.9W	61	194	230	05m42s
0940	2112 Jan 29	03:49:52	99	152	T	-0.9292	1.0346	80.6S	164.3W	21	287	322	01m56s
0941	2113 Jun 13	17:26:00	100	129	T	-0.7097	1.0367	21.7S	74.4W	45	351	174	03m36s
0942	2113 Dec 08	09:03:27	100	134	A	0.5049	0.9296	7.1N	48.4E	60	191	304	09m35s
0943	2114 Jun 03	09:14:09	101	139	T	0.0525	1.0766	25.4N	40.7E	87	167	248	06m32s
0944	2114 Nov 27	08:24:14	101	144	A	-0.1815	0.9223	31.3S	47.8E	79	17	298	09m14s
0945	2115 May 24	02:13:56	101	149	T	0.7912	1.0557	67.8N	108.8E	37	134	301	03m24s
0946	2115 Nov 16	09:58:55	101	154	A	-0.8664	0.9503	68.7S	28.3W	30	63	365	03m32s
0947	2117 Apr 02	06:15:20	102	131	A	-0.4460	0.9333	18.4S	100.5E	63	330	274	07m30s
0948	2117 Sep 26	00:55:42	102	136	T	0.4442	1.0645	21.9N	177.8E	64	211	233	05m03s
0949	2118 Mar 22	06:00:55	103	141	A	0.2719	0.9382	14.3N	84.1E	74	150	237	06m50s
0950	2118 Sep 15	16:28:26	103	146	T	-0.2823	1.0349	11.5S	75.8W	74	29	122	03m04s
0951	2119 Mar 11	10:19:19	103	151	A	0.9693	0.9694	56.7N	29.8W	14	120	451	02m13s
0952	2120 Jul 25	14:40:01	104	128	An	0.9948	0.9343	66.0N	89.8E	4	312	-	04m00s
0953	2121 Jan 19	02:54:15	104	133	T	-0.4190	1.0371	43.9S	149.5E	65	337	137	02m52s
0954	2121 Jul 14	16:42:38	104	138	A	0.2125	0.9758	33.6N	64.9W	78	197	88	02m32s
0955	2122 Jan 08	15:48:51	105	143	A	0.2713	0.9907	6.9S	58.8W	74	168	34	01m02s
0956	2122 Jul 04	01:25:31	105	148	T	-0.5649	1.0280	11.0S	154.0E	56	10	114	02m56s
0957	2122 Dec 28	22:00:56	105	153	A+	1.0072	0.9451	65.3N	170.4W	0	161	-	-
0958	2124 May 14	01:59:10	106	130	T	0.5286	1.0464	50.3N	142.5E	58	167	182	03m34s
0959	2124 Nov 06	06:36:34	106	135	A	-0.5921	0.9724	51.6S	66.2E	53	18	123	02m26s
0960	2125 May 03	13:42:33	106	140	A	-0.2263	0.9915	3.0N	23.3W	77	349	31	00m59s

Cat Num	Calendar Date	TD of Greatest Eclipse	ΔT s	Saros Num	Ecl Type	Gamma	Ecl Mag	Lat °	Long °	Sun Alt °	Sun Azm °	Path Width km	Central Line Dur
0961	2125 Oct 26	17:30:49	107	145	T	0.1461	1.0329	4.5S	84.2W	82	194	112	03m15s
0962	2126 Apr 22	18:04:22	107	150	A-	-1.0051	0.9514	71.1S	40.7W	0	311	-	-
0963	2126 Oct 16	09:12:51	107	155	T	0.8345	1.0534	45.3N	57.9E	33	203	319	04m00s
0964	2128 Mar 01	07:48:32	108	132	A	0.4596	0.9940	18.9N	58.4E	63	163	24	00m37s
0965	2128 Aug 25	23:44:34	108	137	A	-0.3562	0.9694	9.8S	179.3E	69	15	117	03m41s
0966	2129 Feb 18	20:44:37	109	142	T	-0.2526	1.0411	25.6S	123.2W	75	344	142	03m38s
0967	2129 Aug 15	01:33:04	109	147	A	0.4055	0.9442	37.4N	165.1E	66	196	225	06m15s
0968	2130 Feb 08	12:35:23	109	152	T	-0.9212	1.0356	75.9S	51.1E	22	300	313	02m03s
0969	2131 Jun 25	00:43:16	110	129	T	-0.7813	1.0393	28.1S	174.0E	38	356	211	03m43s
0970	2131 Dec 19	17:06:50	110	134	A	0.5165	0.9267	7.6N	73.5W	59	186	321	10m14s
0971	2132 Jun 13	16:46:24	111	139	Tm	-0.0186	1.0788	22.3N	70.9W	89	350	255	06m55s
0972	2132 Dec 07	16:18:43	111	144	A	-0.1661	0.9215	32.2S	68.7W	80	13	301	09m33s
0973	2133 Jun 03	09:45:16	111	149	T	0.7247	1.0567	66.6N	10.0E	43	149	272	03m36s
0974	2133 Nov 26	18:05:55	111	154	A	-0.8473	0.9513	72.0S	144.3W	32	57	337	03m27s
0975	2135 Apr 13	13:27:05	112	131	A	-0.4973	0.9349	17.6S	7.2W	60	332	274	07m30s
0976	2135 Oct 07	09:00:03	113	136	T	0.4884	1.0603	20.3N	56.8E	61	210	224	04m50s
0977	2136 Apr 01	13:26:19	113	141	A	0.2295	0.9430	16.5N	26.8W	77	150	216	06m14s
0978	2136 Sep 26	00:12:14	113	146	T	-0.2309	1.0292	13.0S	168.7E	77	30	101	02m34s
0979	2137 Mar 21	18:16:38	113	151	A	0.9369	0.9769	55.6N	145.5W	20	121	233	01m40s
0980	2139 Jan 30	11:42:25	115	133	T	-0.4255	1.0364	41.0S	19.9E	65	333	135	02m49s
0981	2139 Jul 25	23:26:33	115	138	A	0.2946	0.9778	35.8N	162.7W	73	202	83	02m13s
0982	2140 Jan 20	00:23:11	115	143	A	0.2676	0.9882	5.5S	172.6E	75	163	43	01m17s
0983	2140 Jul 14	08:36:11	116	148	T	-0.4861	1.0322	6.7S	45.7E	61	14	124	03m18s
0984	2141 Jan 08	06:12:38	116	153	A+	1.0024	0.9522	64.3N	57.0E	0	151	-	-
0985	2142 May 25	09:32:37	117	130	T	0.5938	1.0449	57.4N	31.1E	53	171	187	03m17s
0986	2142 Nov 17	14:43:08	117	135	A	-0.6117	0.9727	56.4S	53.2W	52	14	124	02m19s
0987	2143 May 14	20:58:14	117	140	Am	-0.1637	0.9908	9.4N	133.5W	81	352	33	01m05s
0988	2143 Nov 07	01:51:16	118	145	T	0.1206	1.0326	9.4S	150.0E	83	191	111	03m14s
0989	2144 May 03	01:02:06	118	150	A	-0.9441	0.9363	53.6S	176.7W	19	341	727	06m09s
0990	2144 Oct 26	17:32:40	118	155	T	0.8037	1.0512	39.2N	72.0W	36	198	284	04m05s
0991	2146 Mar 12	15:58:15	119	132	A	0.4821	0.9995	24.4N	65.9W	61	161	2	00m03s
0992	2146 Sep 06	06:44:00	120	137	A	-0.4249	0.9640	17.8S	71.7E	65	18	143	04m13s
0993	2147 Mar 02	05:18:54	120	142	T	-0.2359	1.0452	20.5S	107.9E	76	343	155	04m02s
0994	2147 Aug 26	08:09:15	120	147	A	0.3271	0.9425	29.0N	64.4E	71	197	224	06m49s
0995	2148 Feb 19	21:18:00	120	152	T	-0.9111	1.0370	70.9S	89.1W	24	309	305	02m13s
0996	2149 Jul 05	07:59:34	121	129	T	-0.8544	1.0408	36.3S	61.6E	31	0	264	03m38s
0997	2149 Dec 30	01:13:04	122	134	A	0.5253	0.9245	8.6N	163.9E	58	182	334	10m42s
0998	2150 Jun 25	00:17:25	122	139	T	-0.0911	1.0802	18.3N	177.3E	85	356	260	07m14s
0999	2150 Dec 19	00:17:02	122	144	A	-0.1535	0.9211	32.3S	174.1E	81	8	302	09m46s
1000	2151 Jun 14	17:13:45	123	149	T	0.6561	1.0569	63.7N	90.2W	49	163	249	03m48s
1001	2151 Dec 08	02:18:31	123	154	A	-0.8320	0.9526	75.1S	102.3E	33	47	314	03m22s
1002	2153 Apr 23	20:29:24	124	131	A	-0.5558	0.9364	17.9S	112.7W	56	334	279	07m31s
1003	2153 Oct 17	17:12:17	124	136	T	0.5259	1.0560	18.8N	66.6W	58	208	214	04m36s
1004	2154 Apr 12	20:43:01	125	141	A	0.1794	0.9478	18.2N	135.1W	80	152	195	05m42s
1005	2154 Oct 07	08:03:50	125	146	T	-0.1867	1.0234	15.1S	51.3E	79	29	81	02m05s
1006	2155 Apr 02	02:06:34	125	151	A	0.8975	0.9844	55.6N	100.4E	26	123	123	01m07s
1007	2155 Sep 26	17:14:27	126	156	A	-0.9654	0.9593	58.6S	143.9W	15	68	569	02m55s
1008	2157 Feb 09	20:25:36	127	133	T	-0.4358	1.0362	37.7S	109.3W	64	330	135	02m49s
1009	2157 Aug 05	06:14:19	127	138	A	0.3743	0.9792	37.1N	98.7E	68	207	80	01m59s
1010	2158 Jan 30	08:54:37	127	143	A	0.2619	0.9863	3.4S	44.6E	75	160	50	01m27s
1011	2158 Jul 25	15:49:17	128	148	T	-0.4086	1.0356	3.4S	62.7W	66	18	131	03m32s
1012	2159 Jan 19	14:23:26	128	153	A+	0.9974	0.9600	63.4N	75.1W	0	141	-	-
1013	2160 Jun 04	16:58:36	129	130	T	0.6645	1.0428	64.5N	75.8W	48	178	192	02m58s
1014	2160 Nov 27	22:58:32	130	135	A	-0.6247	0.9734	60.1S	172.5W	51	8	123	02m11s
1015	2161 May 25	04:05:43	130	140	A	-0.0950	0.9898	15.7N	118.9E	85	355	36	01m12s
1016	2161 Nov 17	10:19:30	130	145	T	0.1012	1.0325	13.4S	22.7E	84	188	110	03m13s
1017	2162 May 14	07:52:46	131	150	A	-0.8775	0.9396	42.3S	71.8E	28	349	467	06m37s
1018	2162 Nov 07	01:59:40	131	155	T	0.7788	1.0489	34.1N	157.3E	39	193	258	04m05s
1019	2164 Mar 23	00:02:47	132	132	H	0.5096	1.0051	30.4N	171.1E	59	159	20	00m29s
1020	2164 Sep 16	13:48:20	132	137	A	-0.4885	0.9583	25.7S	37.3W	61	19	172	04m42s

Cat Num	Calendar Date	TD of Greatest Eclipse	ΔT s	Saros Num	Ecl Type	Gamma	Ecl Mag	Lat °	Long °	Sun Alt °	Sun Azm °	Path Width km	Central Line Dur
1021	2165 Mar 12	13:45:50	133	142	T	-0.2130	1.0495	14.9S	19.7W	78	342	168	04m27s
1022	2165 Sep 05	14:52:45	133	147	A	0.2549	0.9406	20.7N	38.4W	75	198	227	07m22s
1023	2166 Mar 02	05:53:21	134	152	T	-0.8957	1.0388	65.4S	133.5E	26	317	294	02m26s
1024	2166 Aug 25	16:13:35	134	157	An	0.9900	0.9531	74.4N	40.5E	7	285	-	03m00s
1025	2167 Jul 16	15:17:48	135	129	T	-0.9262	1.0410	46.8S	53.3W	22	6	368	03m19s
1026	2168 Jan 10	09:19:03	135	134	A	0.5337	0.9230	10.3N	41.2E	58	178	344	10m55s
1027	2168 Jul 05	07:45:23	135	139	T	-0.1660	1.0807	13.2N	65.4E	81	0	264	07m26s
1028	2168 Dec 29	08:19:32	136	144	A	-0.1444	0.9215	31.6S	55.8E	82	2	300	09m52s
1029	2169 Jun 25	00:37:09	136	149	T	0.5841	1.0562	59.2N	167.6E	54	173	229	03m58s
1030	2169 Dec 18	10:37:07	136	154	A	-0.8213	0.9544	77.3S	7.0W	34	31	295	03m15s
1031	2171 May 05	03:23:15	138	131	A	-0.6209	0.9378	19.4S	143.8E	51	337	289	07m32s
1032	2171 Oct 29	01:31:03	138	136	T	0.5577	1.0516	17.6N	168.1E	56	206	203	04m23s
1033	2172 Apr 23	03:53:15	138	141	A	0.1234	0.9528	19.2N	118.6E	83	154	174	05m11s
1034	2172 Oct 17	16:01:36	139	146	H3	-0.1484	1.0174	17.3S	67.6W	81	28	60	01m34s
1035	2173 Apr 12	09:49:40	139	151	A	0.8515	0.9919	56.2N	11.3W	31	126	53	00m34s
1036	2173 Oct 07	00:39:13	140	156	A	-0.9186	0.9558	57.8S	113.0E	23	62	402	03m17s
1037	2175 Feb 21	05:04:24	141	133	T	-0.4495	1.0362	34.2S	121.9E	63	328	135	02m50s
1038	2175 Aug 16	13:08:17	141	138	A	0.4497	0.9802	37.6N	1.5W	63	211	78	01m50s
1039	2176 Feb 10	17:21:21	141	143	A	0.2532	0.9849	0.9S	82.3W	75	156	55	01m34s
1040	2176 Aug 04	23:05:55	142	148	T	-0.3332	1.0383	1.3S	171.6W	71	21	136	03m40s
1041	2177 Jan 29	22:30:30	142	153	An	0.9897	0.9212	57.6N	164.1E	7	140	-	06m55s
1042	2178 Jun 16	00:20:42	143	130	T	0.7379	1.0396	71.0N	176.3W	42	190	198	02m36s
1043	2178 Dec 09	07:20:02	144	135	A	-0.6338	0.9745	62.4S	68.9E	50	360	118	02m03s
1044	2179 Jun 05	11:05:36	144	140	A	-0.0209	0.9884	21.5N	13.9E	89	359	41	01m21s
1045	2179 Nov 28	18:54:18	145	145	T	0.0867	1.0325	16.5S	105.7W	85	184	110	03m12s
1046	2180 May 24	14:34:27	145	150	A	-0.8035	0.9422	32.6S	34.0W	36	354	359	06m59s
1047	2180 Nov 17	10:34:01	146	155	T	0.7605	1.0465	30.1N	25.4E	40	189	238	04m03s
1048	2182 Apr 03	07:59:43	147	132	H	0.5440	1.0108	36.9N	49.9E	57	159	44	00m58s
1049	2182 Sep 27	20:58:45	147	137	A	-0.5462	0.9527	33.5S	147.8W	57	21	205	05m05s
1050	2183 Mar 23	22:06:49	148	142	T	-0.1848	1.0540	8.9S	146.2W	79	342	181	04m54s
1051	2183 Sep 16	21:42:37	148	147	A	0.1877	0.9384	12.8N	143.0W	79	198	233	07m53s
1052	2184 Mar 12	14:22:32	148	152	T	-0.8755	1.0409	59.4S	1.3W	29	324	283	02m43s
1053	2184 Sep 04	23:11:00	149	157	A	0.9185	0.9576	67.1N	124.4W	23	227	393	03m12s
1054	2185 Jul 26	22:38:16	150	129	Ts	-0.9967	1.0370	67.9S	179.6W	1	21	-	02m27s
1055	2186 Jan 20	17:23:44	150	134	A	0.5426	0.9221	12.8N	81.4W	57	174	350	10m53s
1056	2186 Jul 16	15:14:54	150	139	T	-0.2396	1.0805	7.4N	47.6W	76	4	267	07m29s
1057	2187 Jan 09	16:23:41	151	144	A	-0.1365	0.9224	30.0S	63.2W	82	358	296	09m51s
1058	2187 Jul 06	07:58:31	151	149	T	0.5108	1.0548	53.6N	62.7E	59	181	211	04m06s
1059	2187 Dec 29	18:59:03	152	154	A	-0.8126	0.9565	77.7S	112.3W	35	10	274	03m07s
1060	2189 May 15	10:08:34	153	131	A	-0.6928	0.9387	22.6S	42.2E	46	340	309	07m31s
1061	2189 Nov 08	09:57:28	153	136	T	0.5831	1.0474	16.5N	40.5E	54	202	192	04m10s
1062	2190 May 04	10:56:30	154	141	A	0.0607	0.9577	19.4N	14.3E	86	157	154	04m45s
1063	2190 Oct 29	00:05:50	154	146	H	-0.1160	1.0116	19.6S	172.1E	83	25	40	01m04s
1064	2191 Apr 23	17:26:06	155	151	A	0.7991	0.9993	57.0N	120.3W	37	130	4	00m03s
1065	2191 Oct 18	08:11:12	155	156	A	-0.8783	0.9516	58.7S	4.1E	28	59	365	03m39s
1066	2193 Mar 03	13:36:08	157	133	T	-0.4689	1.0365	30.9S	5.5W	62	327	137	02m53s
1067	2193 Aug 26	20:09:19	157	138	A	0.5200	0.9806	37.4N	104.0W	58	214	80	01m45s
1068	2194 Feb 21	01:41:31	157	143	A	0.2396	0.9840	1.9N	152.4E	76	154	58	01m38s
1069	2194 Aug 16	06:28:08	158	148	T	-0.2616	1.0403	0.2S	78.4E	75	24	139	03m44s
1070	2195 Feb 10	06:34:27	158	153	An	0.9797	0.9218	55.2N	40.5E	11	136	-	06m52s
1071	2195 Aug 05	22:21:03	159	158	Ts	-0.9843	1.0618	56.1S	165.3E	9	40	-	04m03s
1072	2196 Jun 26	07:37:40	160	130	T	0.8149	1.0356	76.3N	95.8E	35	213	208	02m12s
1073	2196 Dec 19	15:47:09	160	135	A	-0.6388	0.9762	63.1S	49.8W	50	350	111	01m53s
1074	2197 Jun 15	17:59:33	161	140	A	0.0574	0.9864	26.8N	88.7W	87	184	48	01m32s
1075	2197 Dec 09	03:35:07	161	145	T	0.0768	1.0329	18.5S	124.9E	86	180	111	03m13s
1076	2198 Jun 04	21:11:35	162	150	A	-0.7259	0.9443	24.2S	136.9W	43	359	299	07m13s
1077	2198 Nov 28	19:12:46	162	155	T	0.7459	1.0442	26.9N	107.2W	42	184	221	03m58s
1078	2200 Apr 14	15:49:57	163	132	T	0.5847	1.0165	43.8N	69.5W	54	158	69	01m23s
1079	2200 Oct 09	04:16:21	164	137	A	-0.5972	0.9470	41.1S	100.2E	53	22	241	05m25s
1080	2201 Apr 04	06:19:56	164	142	T	-0.1495	1.0584	2.7S	89.0E	81	343	194	05m20s

Cat Num	Calendar Date	TD of Greatest Eclipse	ΔT s	Saros Num	Ecl Type	Gamma	Ecl Mag	Lat °	Long °	Sun Alt °	Sun Azm °	Path Width km	Central Line Dur
1081	2201 Sep 28	04:41:51	165	147	A	0.1281	0.9361	5.2N	110.2E	83	198	240	08m21s
1082	2202 Mar 24	22:42:58	165	152	T	-0.8484	1.0431	52.9S	133.1W	32	330	271	03m03s
1083	2202 Sep 17	06:18:53	166	157	A	0.8545	0.9597	57.1N	113.0E	31	214	281	03m24s
1084	2204 Feb 02	01:25:25	167	134	A	0.5535	0.9218	16.0N	156.7E	56	170	353	10m38s
1085	2204 Jul 27	22:44:32	168	139	T	-0.3129	1.0793	1.0N	161.3W	72	8	269	07m22s
1086	2205 Jan 21	00:27:32	168	144	A	-0.1281	0.9241	27.5S	177.4E	82	353	289	09m42s
1087	2205 Jul 17	15:17:59	169	149	T	0.4366	1.0525	47.2N	44.2W	64	186	193	04m10s
1088	2206 Jan 10	03:24:08	169	154	A	-0.8060	0.9593	75.9S	139.3E	36	351	252	02m57s
1089	2207 May 27	16:47:46	170	131	A	-0.7693	0.9393	27.5S	58.2W	40	343	347	07m25s
1090	2207 Nov 20	18:30:26	171	136	T	0.6028	1.0434	15.8N	89.0W	53	198	180	03m56s
1091	2208 May 15	17:53:06	171	141	A	-0.0080	0.9625	18.7N	88.2W	90	334	136	04m19s
1092	2208 Nov 09	08:17:12	172	146	H	-0.0905	1.0059	21.8S	50.2E	85	22	20	00m34s
1093	2209 May 05	00:56:53	172	151	H	0.7413	1.0066	57.7N	133.2E	42	136	34	00m28s
1094	2209 Oct 29	15:50:20	173	156	A	-0.8445	0.9472	60.7S	107.5W	32	56	358	04m02s
1095	2211 Mar 15	22:01:40	174	133	T	-0.4931	1.0368	27.8S	131.8W	60	327	140	02m57s
1096	2211 Sep 08	03:17:18	175	138	A	0.5855	0.9808	36.9N	151.3E	54	216	83	01m43s
1097	2212 Mar 04	09:55:00	175	143	A	0.2211	0.9834	4.9N	28.8E	77	152	60	01m40s
1098	2212 Aug 27	13:56:17	176	148	T	-0.1939	1.0416	0.1S	32.9W	79	27	142	03m45s
1099	2213 Feb 21	14:30:14	176	153	A	0.9635	0.9230	53.4N	79.9W	15	133	1080	06m44s
1100	2213 Aug 17	05:56:32	177	158	T	-0.9161	1.0653	46.0S	59.1E	23	36	525	04m35s
1101	2214 Jul 08	14:52:45	178	130	T	0.8926	1.0303	78.1N	27.0E	26	253	230	01m46s
1102	2215 Jan 01	00:16:36	178	135	A	-0.6427	0.9783	62.3S	169.2W	50	340	101	01m41s
1103	2215 Jun 28	00:48:45	179	140	A	0.1388	0.9839	31.4N	170.8E	82	189	58	01m44s
1104	2215 Dec 21	12:20:08	179	145	T	0.0701	1.0337	19.5S	5.4W	86	175	114	03m14s
1105	2216 Jun 16	03:41:04	180	150	A	-0.6420	0.9458	16.7S	123.3E	50	3	260	07m20s
1106	2216 Dec 10	03:57:51	181	155	T	0.7367	1.0421	24.8N	119.0E	42	180	208	03m51s
1107	2218 Apr 25	23:33:14	182	132	T	0.6321	1.0219	51.1N	173.1E	51	158	96	01m43s
1108	2218 Oct 20	11:41:55	183	137	A	-0.6411	0.9416	48.4S	13.4W	50	23	280	05m41s
1109	2219 Apr 15	14:26:32	183	142	T	-0.1086	1.0628	3.7N	34.1W	84	344	207	05m45s
1110	2219 Oct 09	11:48:35	184	147	A	0.0744	0.9338	2.0S	1.7E	86	197	248	08m46s
1111	2220 Apr 04	06:56:41	184	152	T	-0.8162	1.0454	46.2S	97.7E	35	335	260	03m26s
1112	2220 Sep 27	13:35:07	185	157	A	0.7966	0.9609	48.0N	4.1W	37	207	232	03m36s
1113	2222 Feb 12	09:23:17	186	134	A	0.5669	0.9220	20.0N	35.4E	55	166	355	10m14s
1114	2222 Aug 08	06:17:05	187	139	T	-0.3837	1.0774	6.0S	83.6E	67	11	270	07m06s
1115	2223 Feb 01	08:29:43	187	144	A	-0.1179	0.9263	24.1S	57.9E	83	349	279	09m26s
1116	2223 Jul 28	22:38:03	188	149	T	0.3636	1.0495	40.2N	153.0W	68	190	176	04m09s
1117	2224 Jan 21	11:48:53	188	154	A	-0.7984	0.9626	72.4S	23.9E	37	339	227	02m46s
1118	2225 Jun 06	23:21:31	190	131	A	-0.8497	0.9392	34.6S	157.8W	32	347	425	07m10s
1119	2225 Dec 01	03:08:36	191	136	T	0.6178	1.0398	15.4N	140.1E	52	194	169	03m43s
1120	2226 May 27	00:45:11	191	141	A	-0.0810	0.9670	16.8N	170.2E	85	344	119	03m55s
1121	2226 Nov 20	16:34:56	192	146	Hm	-0.0711	1.0005	23.7S	73.0W	86	19	2	00m03s
1122	2227 May 16	08:21:31	192	151	T	0.6774	1.0135	57.7N	29.4E	47	144	63	00m59s
1123	2227 Nov 09	23:36:42	193	156	A	-0.8171	0.9429	63.3S	139.3E	35	53	364	04m24s
1124	2229 Mar 26	06:17:34	194	133	T	-0.5251	1.0371	25.5S	104.2E	58	328	144	03m02s
1125	2229 Sep 18	10:34:51	195	138	A	0.6439	0.9805	36.2N	43.4E	50	217	89	01m44s
1126	2230 Mar 15	18:00:25	196	143	A	0.1964	0.9831	7.9N	92.6W	79	151	61	01m40s
1127	2230 Sep 07	21:30:39	196	148	T	-0.1308	1.0424	0.7S	145.8W	82	28	143	03m44s
1128	2231 Mar 04	22:20:23	197	153	A	0.9430	0.9246	52.4N	161.7E	19	130	837	06m32s
1129	2231 Aug 28	13:35:31	197	158	T	-0.8505	1.0661	41.4S	53.5W	31	36	402	04m43s
1130	2232 Jul 18	22:04:56	198	130	T	0.9717	1.0229	72.4N	34.8W	13	299	348	01m14s
1131	2233 Jan 11	08:49:16	199	135	A	-0.6447	0.9811	60.0S	69.0E	50	333	88	01m28s
1132	2233 Jul 08	07:35:24	200	140	A	0.2215	0.9809	35.1N	71.8E	77	194	70	01m59s
1133	2233 Dec 31	21:07:37	200	145	T	0.0649	1.0348	19.5S	136.1W	86	170	117	03m18s
1134	2234 Jun 27	10:09:34	201	150	A	-0.5572	0.9468	10.3S	24.7E	56	8	235	07m18s
1135	2234 Dec 21	12:46:02	201	155	T	0.7299	1.0403	23.5N	15.5W	43	175	197	03m42s
1136	2236 May 06	07:11:02	203	132	T	0.6849	1.0269	58.7N	57.6E	46	159	126	01m59s
1137	2236 Oct 30	19:15:14	204	137	A	-0.6779	0.9365	55.2S	127.8W	47	23	321	05m54s
1138	2237 Apr 25	22:25:03	204	142	T	-0.0606	1.0669	10.1N	155.1W	87	346	219	06m05s
1139	2237 Oct 19	19:06:03	205	147	A	0.0295	0.9316	8.6S	109.0W	88	196	256	09m07s
1140	2238 Apr 15	15:01:45	205	152	T	-0.7771	1.0475	39.3S	28.7W	39	340	250	03m49s

Cat Num	Calendar Date	TD of Greatest Eclipse	ΔT s	Saros Num	Ecl Type	Gamma	Ecl Mag	Lat °	Long °	Sun Alt °	Sun Azm °	Path Width km	Central Line Dur
1141	2238 Oct 08	21:01:18	206	157	A	0.7459	0.9618	40.1N	121.1W	41	202	206	03m47s
1142	2240 Feb 23	17:14:11	208	134	A	0.5859	0.9228	24.7N	84.4W	54	163	356	09m41s
1143	2240 Aug 18	13:52:24	208	139	T	-0.4522	1.0746	13.3S	32.7W	63	14	270	06m40s
1144	2241 Feb 11	16:28:39	209	144	A	-0.1045	0.9292	19.9S	61.4W	84	347	267	09m04s
1145	2241 Aug 08	05:59:21	210	149	T	0.2920	1.0457	32.9N	96.6E	73	193	159	04m02s
1146	2242 Jan 31	20:12:58	210	154	A	-0.7894	0.9665	67.9S	97.2W	38	333	197	02m31s
1147	2243 Jun 18	05:49:56	212	131	A	-0.9342	0.9380	45.6S	103.3E	20	351	652	06m41s
1148	2243 Dec 12	11:52:14	213	136	T	0.6284	1.0365	15.5N	7.6E	51	190	157	03m30s
1149	2244 Jun 06	07:33:12	213	141	Am	-0.1581	0.9712	13.8N	69.3E	81	349	105	03m31s
1150	2244 Dec 01	00:58:17	214	146	A	-0.0568	0.9955	25.1S	162.6E	87	14	16	00m27s
1151	2245 May 26	15:42:04	214	151	T	0.6089	1.0201	56.7N	72.8W	52	153	86	01m30s
1152	2245 Nov 20	07:29:36	215	156	A	-0.7954	0.9387	66.3S	25.7E	37	48	374	04m45s
1153	2247 Apr 06	14:26:51	217	133	T	-0.5624	1.0372	23.8S	18.3W	56	329	149	03m07s
1154	2247 Sep 29	18:01:05	218	138	A	0.6961	0.9801	35.6N	67.3W	46	216	96	01m47s
1155	2248 Mar 26	01:56:01	218	143	A	0.1643	0.9829	10.6N	148.7E	80	151	61	01m41s
1156	2248 Sep 18	05:13:07	219	148	T	-0.0738	1.0426	2.0S	99.2E	86	29	143	03m42s
1157	2249 Mar 15	06:00:45	219	153	A	0.9149	0.9266	52.0N	46.9E	23	128	666	06m18s
1158	2249 Sep 07	21:21:29	220	158	T	-0.7906	1.0656	39.4S	168.9W	38	37	343	04m42s
1159	2251 Jan 22	17:21:41	222	135	A	-0.6481	0.9844	56.9S	54.6W	49	327	72	01m12s
1160	2251 Jul 19	14:18:46	223	140	A	0.3062	0.9773	38.0N	25.7W	72	200	85	02m16s
1161	2252 Jan 12	05:57:04	223	145	T	0.0607	1.0365	18.5S	92.6E	87	165	123	03m23s
1162	2252 Jul 07	16:34:11	224	150	A	-0.4686	0.9473	4.9S	72.1W	62	12	218	07m10s
1163	2252 Dec 31	21:37:06	225	155	T	0.7257	1.0389	23.1N	150.6W	43	170	189	03m33s
1164	2254 May 17	14:43:39	226	132	T	0.7426	1.0315	66.7N	55.5W	42	161	160	02m09s
1165	2254 Nov 11	02:55:16	227	137	A	-0.7087	0.9317	61.4S	117.8E	45	21	363	06m05s
1166	2255 May 07	06:18:06	228	142	T	-0.0076	1.0706	16.4N	85.7E	90	346	230	06m22s
1167	2255 Oct 31	02:32:03	228	147	A	-0.0088	0.9295	14.5S	138.7E	89	11	264	09m24s
1168	2256 Apr 25	22:58:34	229	152	T	-0.7316	1.0495	32.3S	152.4W	43	344	240	04m14s
1169	2256 Oct 19	04:37:31	230	157	A	0.7025	0.9624	33.1N	120.8E	45	198	190	03m59s
1170	2258 Mar 06	00:58:23	232	134	A	0.6101	0.9239	30.2N	157.3E	52	160	358	09m04s
1171	2258 Aug 29	21:33:04	232	139	T	-0.5161	1.0712	20.9S	150.8W	59	17	269	06m09s
1172	2259 Feb 23	00:23:41	233	144	A	-0.0875	0.9326	15.0S	179.7E	85	345	253	08m36s
1173	2259 Aug 19	13:22:17	234	149	T	0.2226	1.0412	25.3N	15.1W	77	195	141	03m49s
1174	2260 Feb 12	04:34:24	234	154	A	-0.7776	0.9711	62.7S	138.7E	39	331	165	02m15s
1175	2261 Dec 22	20:38:49	237	136	T	0.6360	1.0337	16.1N	125.7W	50	185	147	03m17s
1176	2262 Jun 17	14:19:14	238	141	A	-0.2377	0.9750	9.8N	31.7W	76	353	92	03m08s
1177	2262 Dec 12	09:25:02	238	146	A	-0.0460	0.9910	25.8S	37.5E	87	10	32	00m56s
1178	2263 Jun 06	22:58:57	239	151	T	0.5365	1.0261	54.4N	174.6W	57	162	105	02m01s
1179	2263 Dec 01	15:28:44	240	156	A	-0.7794	0.9349	69.2S	87.3W	38	41	388	05m06s
1180	2265 Apr 16	22:26:19	242	133	T	-0.6073	1.0371	23.1S	138.4W	52	331	154	03m11s
1181	2265 Oct 10	01:37:34	243	138	A	0.7405	0.9796	35.1N	178.6E	42	215	105	01m51s
1182	2266 Apr 06	09:42:37	243	143	Am	0.1255	0.9829	12.9N	32.5E	83	151	61	01m42s
1183	2266 Sep 29	13:03:56	244	148	T	-0.0233	1.0425	3.7S	18.0W	89	28	142	03m40s
1184	2267 Mar 26	13:33:45	245	153	A	0.8810	0.9290	52.3N	65.2W	28	128	549	06m03s
1185	2267 Sep 19	05:12:14	245	158	T	-0.7347	1.0642	38.8S	74.4E	42	38	304	04m34s
1186	2269 Feb 02	01:53:05	247	135	A	-0.6529	0.9883	53.2S	179.8W	49	323	54	00m54s
1187	2269 Jul 29	21:03:04	248	140	A	0.3893	0.9732	39.9N	122.8W	67	205	104	02m35s
1188	2270 Jan 22	14:46:28	249	145	T	0.0560	1.0386	16.7S	38.9W	87	161	130	03m29s
1189	2270 Jul 18	22:59:54	250	150	A	-0.3810	0.9474	0.7S	168.5W	68	16	208	06m57s
1190	2271 Jan 12	06:28:08	250	155	T	0.7217	1.0379	23.3N	74.4E	44	165	182	03m25s
1191	2272 May 27	22:11:12	252	132	T	0.8053	1.0353	75.0N	164.8W	36	166	202	02m14s
1192	2272 Nov 21	10:42:52	253	137	A	-0.7327	0.9275	66.8S	4.3E	43	16	402	06m15s
1193	2273 May 17	14:04:30	254	142	Tm	0.0515	1.0738	22.5N	31.3W	87	173	240	06m31s
1194	2273 Nov 10	10:07:17	255	147	A	-0.0399	0.9278	19.6S	24.7E	88	10	272	09m34s
1195	2274 May 07	06:47:36	255	152	T	-0.6799	1.0510	25.4S	86.6E	47	348	230	04m37s
1196	2274 Oct 30	12:24:17	256	157	A	0.6666	0.9629	27.0N	0.8E	48	195	179	04m08s
1197	2276 Mar 16	08:34:02	258	134	A	0.6411	0.9253	36.4N	40.7E	50	158	362	08m23s
1198	2276 Sep 09	05:18:47	259	139	T	-0.5755	1.0671	28.5S	89.6E	55	20	266	05m33s
1199	2277 Mar 05	08:11:55	260	144	A	-0.0644	0.9366	9.5S	62.0E	86	343	236	08m04s
1200	2277 Aug 29	20:49:10	260	149	T	0.1572	1.0362	17.8N	128.3W	81	196	123	03m28s

Cat Num	Calendar Date	TD of Greatest Eclipse	ΔT s	Saros Num	Ecl Type	Gamma	Ecl Mag	Lat °	Long °	Sun Alt °	Sun Azm °	Path Width km	Central Line Dur
1201	2278 Feb 22	12:52:47	261	154	A	-0.7628	0.9762	57.1S	13.2E	40	331	131	01m54s
1202	2278 Aug 19	06:46:22	262	159	A	0.9568	0.9712	75.8N	154.2E	16	257	367	01m53s
1203	2280 Jan 03	05:28:11	264	136	T	0.6414	1.0314	17.2N	100.2E	50	180	138	03m04s
1204	2280 Jun 27	21:03:20	265	141	A	-0.3198	0.9784	4.6N	132.9W	71	357	81	02m45s
1205	2280 Dec 22	17:55:43	266	146	A	-0.0391	0.9870	25.8S	88.5W	88	5	46	01m23s
1206	2281 Jun 17	06:14:41	266	151	T	0.4621	1.0316	50.8N	82.6E	62	170	121	02m32s
1207	2281 Dec 11	23:31:24	267	156	A	-0.7667	0.9316	71.4S	162.0E	40	30	400	05m26s
1208	2283 Apr 28	06:18:21	269	133	T	-0.6581	1.0366	23.6S	103.3E	49	334	160	03m13s
1209	2283 Oct 21	09:23:10	270	138	A	0.7783	0.9790	34.9N	61.6E	39	212	116	01m56s
1210	2284 Apr 16	17:19:22	271	143	A	0.0792	0.9827	14.6N	80.9W	85	153	61	01m45s
1211	2284 Oct 09	21:03:47	272	148	T	0.0205	1.0420	5.7S	137.5W	89	209	140	03m39s
1212	2285 Apr 05	20:55:22	273	153	A	0.8378	0.9315	52.9N	173.1W	33	129	459	05m50s
1213	2285 Sep 29	13:11:37	273	158	T	-0.6859	1.0622	39.6S	44.6W	46	38	275	04m24s
1214	2287 Feb 13	10:21:24	276	135	A	-0.6613	0.9926	49.4S	54.6E	48	321	34	00m35s
1215	2287 Aug 10	03:47:41	276	140	A	0.4714	0.9686	41.0N	140.1E	62	210	127	02m56s
1216	2288 Feb 02	23:33:46	277	145	T	0.0492	1.0412	14.2S	170.1W	87	157	138	03m39s
1217	2288 Jul 29	05:25:22	278	150	A	-0.2929	0.9469	2.5N	95.7E	73	19	203	06m46s
1218	2289 Jan 22	15:19:24	279	155	T	0.7181	1.0374	24.3N	60.6W	44	161	178	03m18s
1219	2290 Jun 08	05:35:48	281	132	T	0.8713	1.0382	83.8N	99.2E	29	182	265	02m14s
1220	2290 Dec 02	18:36:40	282	137	A	-0.7515	0.9237	70.9S	106.4W	41	7	439	06m23s
1221	2291 May 28	21:45:28	283	142	T	0.1153	1.0764	28.3N	146.2W	83	176	249	06m34s
1222	2291 Nov 21	17:50:53	284	147	A	-0.0644	0.9263	23.7S	90.6W	86	7	278	09m41s
1223	2292 May 17	14:29:32	284	152	T	-0.6224	1.0521	18.8S	32.0W	51	353	220	04m57s
1224	2292 Nov 09	20:20:06	285	157	A	0.6375	0.9635	22.0N	120.8W	50	191	171	04m14s
1225	2294 Mar 27	16:02:22	288	134	A	0.6777	0.9269	43.2N	74.3W	47	156	370	07m42s
1226	2294 Sep 20	13:09:57	288	139	T	-0.6301	1.0627	36.2S	31.7W	51	22	263	04m56s
1227	2295 Mar 16	15:54:33	289	144	A	-0.0362	0.9409	3.6S	54.7W	88	343	219	07m29s
1228	2295 Sep 10	04:20:19	290	149	Tm	0.0963	1.0307	10.3N	117.1E	84	197	104	03m01s
1229	2296 Mar 04	21:04:45	291	154	A	-0.7417	0.9819	51.1S	112.1W	42	333	95	01m31s
1230	2296 Aug 29	13:45:39	292	159	A	0.8888	0.9689	66.6N	13.2E	27	223	245	02m20s
1231	2298 Jan 13	14:16:26	294	136	T	0.6474	1.0296	19.0N	33.7W	50	176	131	02m52s
1232	2298 Jul 09	03:49:01	295	141	A	-0.4013	0.9811	1.4S	124.8E	66	2	73	02m23s
1233	2299 Jan 03	02:27:42	296	146	A	-0.0341	0.9836	24.9S	145.2E	88	0	58	01m47s
1234	2299 Jun 28	13:27:43	297	151	T	0.3845	1.0365	46.0N	21.2W	67	176	133	03m03s
1235	2299 Dec 23	07:38:41	297	156	A	-0.7583	0.9288	72.5S	53.0E	40	16	412	05m45s
1236	2301 May 09	14:00:58	300	133	T	-0.7161	1.0354	25.5S	12.8W	44	337	168	03m10s
1237	2301 Nov 01	17:19:32	301	138	A	0.8080	0.9786	34.8N	58.9W	36	209	126	02m01s
1238	2302 Apr 29	00:47:19	302	143	A	0.0262	0.9825	15.6N	168.2E	88	157	62	01m49s
1239	2302 Oct 22	05:11:15	302	148	T	0.0584	1.0413	7.8S	101.1E	87	207	139	03m38s
1240	2303 Apr 18	04:09:26	303	153	A	0.7888	0.9341	53.8N	82.0E	38	132	393	05m38s
1241	2303 Oct 11	21:17:24	304	158	T	-0.6424	1.0596	41.1S	164.9W	50	38	252	04m12s
1242	2305 Feb 24	18:46:09	307	135	A	-0.6732	0.9973	45.7S	71.0W	47	320	13	00m13s
1243	2305 Aug 21	10:35:44	308	140	A	0.5498	0.9637	41.5N	41.9E	56	214	155	03m21s
1244	2306 Feb 14	08:17:48	308	145	T	0.0394	1.0441	11.3S	59.3E	88	154	147	03m49s
1245	2306 Aug 10	11:55:09	309	150	A	-0.2083	0.9461	4.6N	0.7W	78	23	202	06m37s
1246	2307 Feb 04	00:08:00	310	155	T	0.7125	1.0373	25.7N	165.1E	44	156	176	03m12s
1247	2307 Jul 30	13:31:15	311	160	A	-0.9574	0.9602	50.0S	50.5W	16	30	500	03m37s
1248	2308 Jun 19	12:57:52	313	132	T	0.9402	1.0396	84.1N	118.8E	19	313	401	02m08s
1249	2308 Dec 14	02:34:52	314	137	A	-0.7662	0.9207	73.4S	146.8E	40	353	470	06m31s
1250	2309 Jun 09	05:21:54	314	142	T	0.1833	1.0783	33.6N	100.9E	79	181	257	06m30s
1251	2309 Dec 03	01:42:04	315	147	A	-0.0833	0.9254	26.9S	152.8E	85	3	282	09m40s
1252	2310 May 29	22:04:49	316	152	T	-0.5599	1.0526	12.5S	148.3W	56	357	210	05m10s
1253	2310 Nov 22	04:24:19	317	157	A	0.6145	0.9642	17.9N	116.0E	52	187	164	04m16s
1254	2312 Apr 07	23:19:31	320	134	A	0.7231	0.9286	50.8N	172.9E	43	153	385	07m00s
1255	2312 Oct 01	21:08:25	321	139	T	-0.6783	1.0578	43.8S	154.8W	47	24	258	04m20s
1256	2313 Mar 27	23:29:30	321	144	A	-0.0010	0.9457	2.6N	169.7W	90	335	200	06m49s
1257	2313 Sep 21	11:56:59	322	149	T	0.0405	1.0249	3.0N	1.2E	88	198	85	02m30s
1258	2314 Mar 17	05:11:54	323	154	A	-0.7159	0.9880	44.9S	123.3E	44	335	60	01m03s
1259	2314 Sep 10	20:49:10	324	159	A	0.8247	0.9654	56.8N	105.1W	34	212	220	02m54s
1260	2316 Jan 25	23:05:16	327	136	T	0.6527	1.0282	21.4N	167.8W	49	172	126	02m42s

Cat Num	Calendar Date	TD of Greatest Eclipse	ΔT s	Saros Num	Ecl Type	Gamma	Ecl Mag	Lat °	Long °	Sun Alt °	Sun Azm °	Path Width km	Central Line Dur
1261	2316 Jul 20	10:36:17	328	141	A	-0.4819	0.9834	8.2S	21.3E	61	5	67	02m02s
1262	2317 Jan 14	10:59:37	329	146	A	-0.0298	0.9807	23.2S	18.6E	88	356	69	02m08s
1263	2317 Jul 09	20:42:40	330	151	T	0.3078	1.0406	40.4N	127.1W	72	182	143	03m32s
1264	2318 Jan 03	15:47:14	331	156	A	-0.7518	0.9265	71.9S	55.5W	41	1	422	06m02s
1265	2319 May 20	21:37:22	333	133	T	-0.7786	1.0336	29.0S	127.6W	39	340	178	03m02s
1266	2319 Nov 13	01:24:38	334	138	A	0.8314	0.9784	35.0N	177.8E	33	205	136	02m04s
1267	2320 May 09	08:04:33	335	143	A	-0.0347	0.9820	15.6N	60.2E	88	337	64	01m56s
1268	2320 Nov 01	13:28:18	336	148	Tm	0.0888	1.0406	9.8S	22.7W	85	204	137	03m38s
1269	2321 Apr 28	11:12:58	337	153	A	0.7314	0.9367	54.5N	18.9W	43	136	341	05m30s
1270	2321 Oct 22	05:31:17	338	158	T	-0.6059	1.0567	43.3S	73.0E	52	37	233	04m00s
1271	2323 Mar 08	03:05:09	341	135	H	-0.6906	1.0023	42.4S	164.2E	46	320	11	00m11s
1272	2323 Sep 01	17:26:08	342	140	A	0.6253	0.9584	41.7N	57.1W	51	218	191	03m48s
1273	2324 Feb 25	16:57:31	342	145	Tm	0.0256	1.0475	8.1S	70.5W	89	152	158	04m02s
1274	2324 Aug 20	18:28:21	343	150	A	-0.1260	0.9449	5.7N	97.9W	83	25	205	06m33s
1275	2325 Feb 14	08:52:35	344	155	T	0.7038	1.0378	27.5N	32.0E	45	152	175	03m08s
1276	2325 Aug 09	20:16:23	345	160	A	-0.8749	0.9648	40.3S	148.0W	29	30	256	03m24s
1277	2326 Dec 25	10:36:52	348	137	A	-0.7774	0.9182	73.6S	41.4E	39	337	496	06m39s
1278	2327 Jun 20	12:55:00	349	142	T	0.2542	1.0795	38.3N	10.2W	75	186	265	06m21s
1279	2327 Dec 14	09:39:46	350	147	A	-0.0969	0.9250	28.8S	35.1E	84	358	284	09m34s
1280	2328 Jun 09	05:33:52	351	152	T	-0.4927	1.0524	6.7S	97.6E	60	1	199	05m16s
1281	2328 Dec 02	12:36:36	352	157	A	0.5974	0.9652	14.8N	8.8W	53	183	157	04m13s
1282	2330 Apr 19	06:29:24	355	134	A	0.7742	0.9302	59.0N	61.0E	39	151	412	06m19s
1283	2330 Oct 13	05:13:40	356	139	T	-0.7208	1.0528	51.2S	80.6E	44	27	251	03m46s
1284	2331 Apr 08	06:57:08	357	144	A	0.0408	0.9506	9.2N	77.1E	88	164	181	06m07s
1285	2331 Oct 02	19:39:15	358	149	T	-0.0097	1.0188	4.0S	116.1W	89	17	64	01m55s
1286	2332 Mar 27	13:11:33	359	154	A	-0.6831	0.9945	38.3S	0.1E	47	338	26	00m30s
1287	2332 Sep 21	03:59:09	360	159	A	0.7666	0.9613	47.9N	140.4E	40	207	216	03m34s
1288	2334 Feb 05	07:50:29	363	136	T	0.6603	1.0272	24.6N	58.8E	49	168	123	02m33s
1289	2334 Jul 31	17:26:32	364	141	A	-0.5609	0.9851	15.6S	83.7W	56	9	64	01m45s
1290	2335 Jan 25	19:29:43	365	146	A	-0.0247	0.9784	20.6S	107.8W	88	352	77	02m25s
1291	2335 Jul 21	03:57:48	366	151	T	0.2306	1.0440	34.0N	125.4E	76	186	151	03m58s
1292	2336 Jan 14	23:56:41	367	156	A	-0.7462	0.9250	69.6S	166.8W	41	349	427	06m18s
1293	2336 Jul 09	19:58:21	368	161	T	0.9598	1.0657	83.2N	47.5E	16	345	799	03m18s
1294	2337 May 31	05:05:56	370	133	T	-0.8470	1.0309	34.6S	119.2E	32	344	195	02m46s
1295	2337 Nov 23	09:37:55	371	138	A	0.8488	0.9786	35.5N	51.8E	32	200	142	02m05s
1296	2338 May 20	15:14:19	372	143	A	-0.1011	0.9812	14.5N	45.9W	84	342	67	02m07s
1297	2338 Nov 12	21:52:53	373	148	T	0.1132	1.0399	11.7S	148.3W	84	201	134	03m38s
1298	2339 May 09	18:08:03	374	153	A	0.6672	0.9392	54.7N	116.5W	48	143	300	05m24s
1299	2339 Nov 02	13:51:49	375	158	T	-0.5750	1.0536	45.8S	50.2W	55	34	215	03m47s
1300	2341 Mar 18	11:18:19	378	135	H	-0.7137	1.0075	39.8S	40.7E	44	321	36	00m36s
1301	2341 Sep 12	00:22:46	379	140	A	0.6950	0.9529	41.7N	158.4W	46	220	234	04m19s
1302	2342 Mar 08	01:32:13	380	145	T	0.0072	1.0511	4.9S	161.0E	90	149	169	04m16s
1303	2342 Sep 01	01:06:54	381	150	A	-0.0480	0.9434	6.1N	163.8E	87	28	209	06m34s
1304	2343 Feb 25	17:32:17	382	155	T	0.6913	1.0385	29.6N	99.6W	46	149	175	03m06s
1305	2343 Aug 21	03:07:04	383	160	A	-0.7956	0.9679	35.1S	110.9E	37	31	186	03m09s
1306	2345 Jan 04	18:40:22	386	137	A	-0.7872	0.9165	71.9S	66.6W	38	323	517	06m45s
1307	2345 Jun 30	20:26:17	387	142	T	0.3267	1.0797	42.1N	119.7W	71	192	272	06m07s
1308	2345 Dec 24	17:41:03	388	147	Am	-0.1082	0.9252	29.7S	83.1W	84	353	284	09m21s
1309	2346 Jun 20	12:58:43	389	152	T	-0.4224	1.0515	1.5S	14.7W	65	5	188	05m12s
1310	2346 Dec 13	20:55:35	390	157	A	0.5848	0.9665	12.8N	135.1W	54	178	149	04m04s
1311	2348 Apr 29	13:29:00	393	134	A	0.8338	0.9315	68.1N	50.8W	33	145	466	05m40s
1312	2348 Oct 23	13:26:55	394	139	T	-0.7564	1.0476	58.2S	45.6W	41	28	242	03m14s
1313	2349 Apr 18	14:16:51	395	144	A	0.0900	0.9557	16.0N	34.1W	85	165	162	05m23s
1314	2349 Oct 13	03:28:53	396	149	H	-0.0532	1.0126	10.6S	125.2E	87	16	43	01m18s
1315	2350 Apr 07	21:06:02	397	154	H	-0.6452	1.0011	31.7S	121.7W	50	340	5	00m06s
1316	2350 Oct 02	11:14:06	399	159	A	0.7131	0.9568	39.8N	26.7E	44	203	222	04m21s
1317	2352 Feb 16	16:32:05	402	136	T	0.6709	1.0266	28.5N	73.8W	48	164	121	02m25s
1318	2352 Aug 11	00:21:34	403	141	A	-0.6366	0.9862	23.6S	169.2E	50	13	63	01m32s
1319	2353 Feb 05	03:56:54	404	146	A	-0.0179	0.9766	17.1S	126.0E	89	349	84	02m38s
1320	2353 Jul 31	11:17:05	405	151	T	0.1558	1.0467	27.2N	15.8E	81	190	158	04m20s

Cat Num	Calendar Date	TD of Greatest Eclipse	ΔT s	Saros Num	Ecl Type	Gamma	Ecl Mag	Lat °	Long °	Sun Alt °	Sun Azm °	Path Width km	Central Line Dur
1321	2354 Jan 25	08:03:19	406	156	A	-0.7388	0.9240	66.0S	78.5E	42	341	427	06m35s
1322	2354 Jul 21	03:28:21	407	161	T	0.8870	1.0697	81.4N	169.6E	27	221	499	03m51s
1323	2355 Jun 11	12:28:17	409	133	T	-0.9196	1.0269	43.3S	7.2E	23	348	233	02m18s
1324	2355 Dec 04	17:58:36	410	138	A	0.8609	0.9792	36.0N	76.4W	30	195	145	02m02s
1325	2356 May 30	22:15:17	411	143	A	-0.1735	0.9800	12.2N	150.0W	80	346	72	02m21s
1326	2356 Nov 23	06:24:54	412	148	T	0.1317	1.0394	13.2S	84.3E	83	197	133	03m40s
1327	2357 May 20	00:54:22	414	153	A	0.5960	0.9415	53.9N	148.9E	53	150	268	05m24s
1328	2357 Nov 12	22:20:22	415	158	T	-0.5514	1.0505	48.4S	174.7W	56	31	200	03m35s
1329	2359 Mar 29	19:24:45	418	135	T	-0.7430	1.0128	37.9S	81.4W	42	323	65	01m02s
1330	2359 Sep 23	07:24:41	419	140	A	0.7595	0.9471	41.9N	98.5E	40	221	291	04m53s
1331	2360 Mar 18	09:59:21	420	145	T	-0.0178	1.0549	1.8S	34.3E	89	331	181	04m33s
1332	2360 Sep 11	07:52:24	421	150	Am	0.0244	0.9415	5.7N	63.6E	89	208	217	06m41s
1333	2361 Mar 08	02:05:55	422	155	T	0.6742	1.0396	31.9N	130.7E	47	146	176	03m06s
1334	2361 Aug 31	10:04:30	424	160	A	-0.7211	0.9701	32.2S	7.7E	44	33	151	02m54s
1335	2363 Jan 16	02:45:06	427	137	A	-0.7955	0.9154	68.8S	179.6W	37	314	532	06m52s
1336	2363 Jul 12	03:55:03	428	142	T	0.4013	1.0792	45.0N	132.4E	66	198	279	05m51s
1337	2364 Jan 05	01:46:47	429	147	A	-0.1161	0.9259	29.4S	157.6E	83	348	281	09m03s
1338	2364 Jun 30	20:19:47	430	152	T	-0.3493	1.0499	2.9N	125.4W	70	9	176	05m00s
1339	2364 Dec 24	05:18:58	431	157	A	0.5752	0.9683	11.6N	97.8E	55	174	139	03m48s
1340	2366 May 10	20:22:07	435	134	A	0.8981	0.9323	77.9N	171.6W	26	129	583	05m03s
1341	2366 Nov 03	21:46:04	436	139	T	-0.7868	1.0426	64.8S	172.3W	38	29	231	02m46s
1342	2367 Apr 29	21:30:02	437	144	Am	0.1452	0.9607	22.8N	143.3W	82	167	144	04m38s
1343	2367 Oct 24	11:25:03	438	149	H	-0.0902	1.0065	16.7S	5.2E	85	15	22	00m40s
1344	2368 Apr 18	04:51:37	439	154	H	-0.5992	1.0079	24.8S	118.7E	53	344	34	00m47s
1345	2368 Oct 12	18:37:19	441	159	A	0.6672	0.9522	32.5N	87.9W	48	199	233	05m13s
1346	2370 Feb 27	01:07:02	444	136	T	0.6865	1.0262	33.2N	155.0E	46	161	122	02m17s
1347	2370 Aug 22	07:22:20	445	141	A	-0.7082	0.9867	32.0S	59.9E	45	17	66	01m22s
1348	2371 Feb 16	12:18:48	446	146	A	-0.0075	0.9753	12.9S	0.6E	89	348	88	02m48s
1349	2371 Aug 11	18:38:03	447	151	T	0.0820	1.0487	19.9N	95.1W	85	192	162	04m36s
1350	2372 Feb 05	16:07:47	449	156	A	-0.7301	0.9237	61.5S	39.0W	43	336	422	06m50s
1351	2372 Jul 31	10:58:29	450	161	T	0.8144	1.0717	71.0N	43.4E	35	209	404	04m18s
1352	2373 Jun 21	19:45:28	452	133	Ts	-0.9954	1.0191	62.7S	102.2W	3	349	-	01m24s
1353	2373 Dec 15	02:25:54	453	138	A	0.8678	0.9803	36.7N	153.3E	29	190	141	01m56s
1354	2374 Jun 11	05:09:55	454	143	A	-0.2505	0.9784	8.8N	107.0E	76	351	79	02m39s
1355	2374 Dec 04	15:02:55	456	148	T	0.1455	1.0390	14.1S	44.5W	82	193	132	03m42s
1356	2375 May 31	07:34:32	457	153	A	0.5200	0.9436	52.0N	55.8E	58	158	243	05m26s
1357	2375 Nov 24	06:54:53	458	158	T	-0.5327	1.0474	50.7S	60.2E	58	26	186	03m23s
1358	2377 Apr 09	03:25:10	461	135	T	-0.7780	1.0180	37.1S	158.1E	39	325	96	01m28s
1359	2377 Oct 03	14:33:16	463	140	A	0.8178	0.9413	42.6N	6.8W	35	220	366	05m29s
1360	2378 Mar 29	18:20:22	464	145	T	-0.0480	1.0587	1.1N	90.7W	87	331	193	04m51s
1361	2378 Sep 22	14:45:47	465	150	A	0.0905	0.9396	4.8N	38.7W	85	209	225	06m54s
1362	2379 Mar 19	10:31:46	466	155	T	0.6511	1.0409	34.3N	3.5E	49	144	177	03m07s
1363	2379 Sep 11	17:09:31	468	160	A	-0.6517	0.9717	30.9S	97.5W	49	34	130	02m42s
1364	2381 Jan 26	10:46:37	471	137	A	-0.8064	0.9149	65.3S	64.7E	36	309	546	06m57s
1365	2381 Jul 22	11:25:01	472	142	T	0.4749	1.0777	46.9N	24.8E	61	205	285	05m32s
1366	2382 Jan 15	09:53:21	474	147	A	-0.1242	0.9274	28.1S	38.0E	83	343	275	08m40s
1367	2382 Jul 12	03:37:51	475	152	T	-0.2743	1.0475	6.5N	125.3E	74	13	164	04m41s
1368	2383 Jan 04	13:46:25	476	157	A	0.5682	0.9706	11.4N	30.3W	55	169	127	03m26s
1369	2384 May 21	03:05:25	480	134	A	0.9702	0.9317	80.8N	3.0W	13	40	1116	04m28s
1370	2384 Nov 14	06:13:19	481	139	T	-0.8102	1.0377	70.9S	61.3E	36	28	217	02m22s
1371	2385 May 10	04:36:48	482	144	A	0.2063	0.9657	29.5N	109.7E	78	169	126	03m53s
1372	2385 Nov 03	19:27:29	483	149	H	-0.1213	1.0004	22.1S	115.6W	83	13	2	00m03s
1373	2386 Apr 29	12:32:24	485	154	H2	-0.5483	1.0147	18.1S	0.7E	57	347	60	01m30s
1374	2386 Oct 24	02:06:42	486	159	A	0.6267	0.9475	26.1N	156.6E	51	196	246	06m09s
1375	2388 Mar 09	09:36:20	490	136	T	0.7065	1.0260	38.5N	24.9E	45	158	124	02m10s
1376	2388 Sep 01	14:30:25	491	141	A	-0.7744	0.9867	40.7S	52.3W	39	22	73	01m15s
1377	2389 Feb 26	20:33:51	492	146	A	0.0078	0.9744	8.1S	123.5W	90	162	92	02m55s
1378	2389 Aug 22	02:05:53	493	151	T	0.0133	1.0500	12.5N	151.7E	89	193	166	04m45s
1379	2390 Feb 16	00:06:58	495	156	A	-0.7177	0.9239	56.4S	157.8W	44	335	411	07m06s
1380	2390 Aug 11	18:31:26	496	161	T	0.7440	1.0724	61.3N	74.7W	42	206	353	04m41s

Cat Num	Calendar Date	TD of Greatest Eclipse	ΔT s	Saros Num	Ecl Type	Gamma	Ecl Mag	Lat °	Long °	Sun Alt °	Sun Azm °	Path Width km	Central Line Dur
1381	2391 Dec 26	10:57:14	500	138	A	0.8724	0.9820	37.6N	21.8E	29	184	131	01m46s
1382	2392 Jun 21	11:57:57	501	143	A	-0.3319	0.9762	4.1N	5.1E	71	355	90	03m02s
1383	2392 Dec 14	23:46:25	502	148	T	0.1550	1.0391	14.5S	174.6W	81	188	133	03m46s
1384	2393 Jun 10	14:08:40	504	153	A	0.4389	0.9453	48.7N	36.6W	64	166	224	05m34s
1385	2393 Dec 04	15:34:34	505	158	T	-0.5188	1.0445	52.6S	65.2W	59	20	174	03m13s
1386	2395 Apr 20	11:17:14	509	135	T	-0.8204	1.0230	37.7S	39.7E	35	327	134	01m52s
1387	2395 Oct 14	21:49:15	510	140	A	0.8692	0.9354	44.0N	114.6W	29	219	471	06m07s
1388	2396 Apr 09	02:33:16	511	145	T	-0.0851	1.0625	3.4N	146.5E	85	332	206	05m12s
1389	2396 Oct 02	21:48:06	512	150	A	0.1494	0.9375	3.5N	143.4W	81	209	234	07m12s
1390	2397 Mar 29	18:49:51	514	155	T	0.6221	1.0423	36.7N	121.1W	51	144	178	03m11s
1391	2397 Sep 22	00:23:54	515	160	A	-0.5891	0.9728	30.9S	155.0E	54	34	118	02m34s
1392	2399 Feb 06	18:46:43	519	137	A	-0.8180	0.9150	61.6S	53.2W	35	307	557	07m01s
1393	2399 Aug 02	18:55:13	520	142	T	0.5482	1.0754	48.0N	82.6W	57	211	291	05m14s
1394	2400 Jan 26	18:00:09	522	147	A	-0.1322	0.9295	26.0S	82.1W	82	339	267	08m13s
1395	2400 Jul 22	10:54:47	523	152	T	-0.1992	1.0444	9.1N	16.8E	79	17	151	04m17s

References

Espenak, F., *Thousand Year Canon of Solar Eclipses: 1501 to 2500*, Astropixels Pub., Portal, Arizona (2014).

Explanatory Supplement to the Ephemeris, H.M. Almanac Office, London (1974).

Meeus, Grosjean, C.C., and Vanderleen, W., *Canon of Solar Eclipses*, Pergamon Press, Oxford, United Kingdom (1966).

Meeus, J., *Mathematical Astronomy Morsels*, Willmann-Bell, pp. 56–62 (1997).

——, *More Mathematical Astronomy Morsels*, Willmann-Bell, pp. 120–126 (2002a).

——, *More Mathematical Astronomy Morsels*, Willmann-Bell, pp. 70–72 (2002b).

——, *"The maximum possible duration of a total solar eclipse,"* J. Br. Astron. Assoc., 113(6) (2003).

——, *Mathematical Astronomy Morsels III*, Willmann-Bell, pp. 109–111, (2004).

——, *Mathematical Astronomy Morsels IV*, Willmann-Bell, pp. 44–45, (2007).

Morrison, L., and Stephenson, F.R., "Historical Values of the Earth's Clock Error ΔT and the Calculation of Eclipses," *J. Hist. Astron.*, Vol. 35 Part 3, August 2004, No. 120, pp, 327–336 (2004).

Morrison LV, Stephenson FR, Hohenkerk CY, Zawilski M., 2021 Addendum 2020 to *'Measurement of the Earth's rotation: 720 BC to AD 2015'*. Proc. R. Soc. A477:20200776. *https://doi.org/10.1098/rspa.2020.0776* (2021)

Stephenson, F.R., *Historical Eclipses and Earth's Rotation*, Cambridge University Press, Cambridge (1997).

van den Bergh, *Periodicity and Variation of Solar (and Lunar) Eclipses*, Tjeenk Willink, and Haarlem, Netherlands (1955).

Eclipse Guides from AstroPixels Publishing

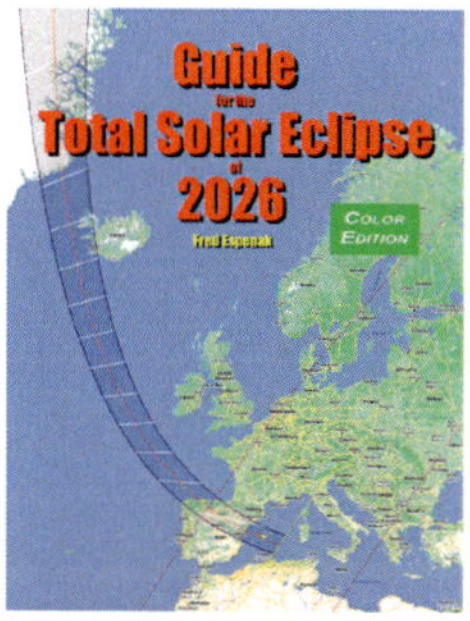

Guide for the Total Solar Eclipse of 2026

The total solar eclipse of August 12, 2026 will be visible from a path across northern Russia, the Arctic, Greenland, Iceland, and Spain. The **Guide for the Total Solar Eclipse of 2026** contains 14 full-page maps of the path. The large scale shows roads, towns and cities, rivers and lakes.

Tables for hundreds of cities provide times each eclipse phase of the along with the eclipse magnitude, duration, and the Sun's altitude. A climatology study identifies areas along the eclipse path where the highest probability of favorable weather may be found.

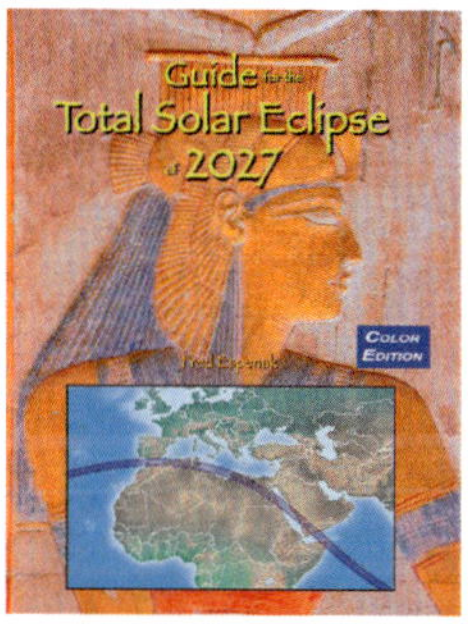

Guide for the Total Solar Eclipse of 2027

The total solar eclipse of August 2, 2027 will be visible from a path across Spain, Morocco, Algeria, Tunisia, Libya, Egypt, Sudan, Saudia Arabia, Yemen, and Somalia. The **Guide for the Total Solar Eclipse of 2027** contains 20 full-page maps of the path.

Tables for hundreds of cities provide times for each eclipse phase along with the eclipse magnitude, duration and the Sun's altitude. Cloud statistics identify areas where the highest probability of clear sky may be found.

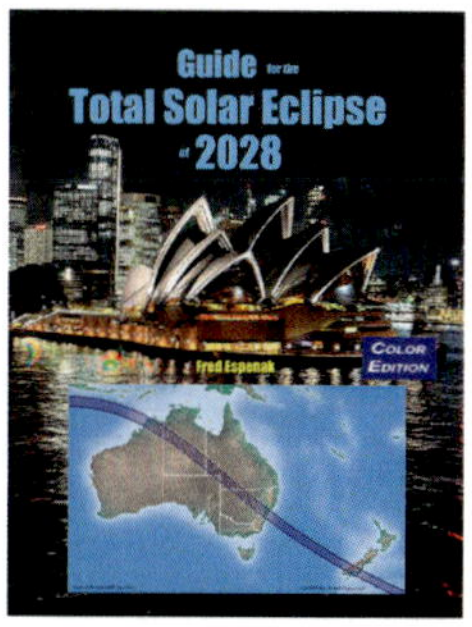

Guide for the Total Solar Eclipse of 2028

The total solar eclipse of July 22, 2028 will be visible from a path across Australia and New Zealand. The **Guide for the Total Solar Eclipse of 2028** contains 19 full-page maps of the path of totality.

Tables for over one hundred cities provide times for each eclipse phase with the duration of totality and the Sun's altitude. Cloud statistics identify areas where the highest probability of clear sky may be found. A preview is given for total eclipses through 2039.

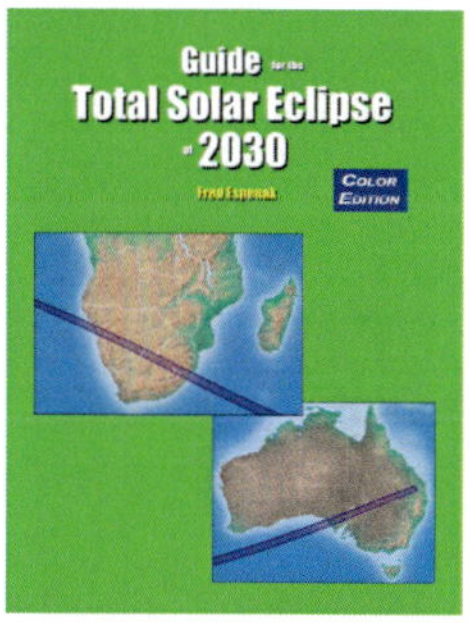

Guide for the Total Solar Eclipse of 2030

The total solar eclipse of November 25, 2030 will be visible from a path across Africa and Australia. The **Guide for the Total Solar Eclipse of 2028** contains 17 full-page maps of the path of totality.

Tables for over one hundred cities provide times for each eclipse phase with the duration of totality and the Sun's altitude. Cloud statistics identify areas where the highest probability of clear sky may be found. A preview is given for total eclipses through 2039.

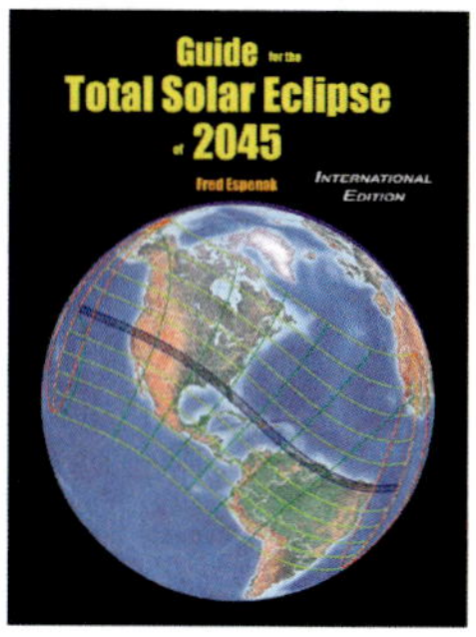

Guide for the Total Solar Eclipse of 2045

The total solar eclipse of August 12, 2045 will be visible from a path across the USA, the Caribbean, and South America. The **Guide for the Total Solar Eclipse of 2045** contains 37 full-page maps of the path of totality.

Tables for hundreds cities provide times for each eclipse phase with the duration of totality, eclipse magnitude, and the Sun's altitude. Cloud statistics identify areas where the highest probability of clear sky may be found.

For more information on each *Guide*, including sample pages of each, visit:

astropixels.com/pubs/guides.html

The *Guides* may be ordered from ***Amazon***

Eclipse Books from AstroPixels Publishing

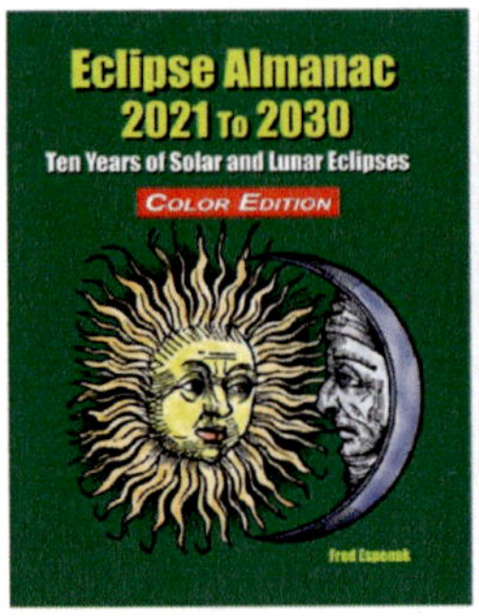

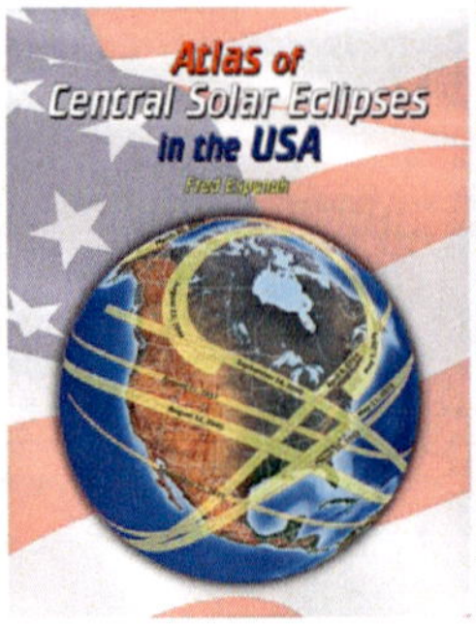

Eclipse Almanac - 2021 to 2030

A concise reference for every eclipse of the Sun and the Moon over a 10-year period. It identifies when and where each event will be seen with maps for each. Additional details about each eclipse are included. Other volumes in this series include the decades 2032 to 2040, 2041 to 2050, 2051 to 2060, and 2061 to 2070.

21st Century Canon of Solar Eclipses

The complete guide to every solar eclipse occurring from 2001 tom 2100 (224 eclipses in all). It includes information and maps for all total, annular, hybrid, and partial eclipses. A special world atlas shows detailed, full page maps of all central eclipse paths (total, annular, and hybrid).

Thousand Year Canon of Solar Eclipses — 1501 to 2500

Contains maps and data for each of the 2,389 solar eclipses occurring over the ten-century period centered on the present era. The maps are arranged twelve to a page at an image scale allowing the determination of eclipse visibility from any location on Earth. A comprehensive catalog lists the essential characteristics of each eclipse.

Atlas of Central Solar Eclipses in the USA

When was the last total eclipse through the USA and when is the next? How often do they happen? What total eclipse tracks passed across the USA during the 17th, 18th, and 19th centuries, etc., and what states did they include? And how often is a total solar eclipse visible from each of the 50 states? The Atlas of Central Solar Eclipses in the USA answers all these questions and more, with hundreds of maps and tables.

21st Century Canon of Lunar Eclipses

The complete guide to every lunar eclipse occurring from 2001 tom 2100 (228 eclipses in all). It includes information and maps for all total, partial, and penumbral eclipses. The predictions use a new model for Earth's elliptical shadows.

Thousand Year Canon of Lunar Eclipses — 1501 to 2500

Contains maps and data for each of the 2,424 lunar eclipses occurring over the ten-century period centered on the present era. The maps are arranged twelve to a page at an image scale allowing the determination of eclipse visibility from any location on Earth. A comprehensive catalog lists the essential characteristics of each eclipse.

All books are available in two editions: 1) Black and White, and 2) Color.

For more information including sample pages of each, visit:

astropixels.com/pubs/guides.html

Each *Book* may be ordered from ***Amazon***

Made in United States
Orlando, FL
04 November 2026

90201773R00079